WATER OF THE SKY

The MIT Press's publishing mission benefits from the generosity of our donors, including Cynthia and John Reed.

WATER OF THE SKY

A DICTIONARY OF 2,000 JAPANESE RAIN WORDS

MIYA ANDO

WITH JOAN HALIFAX

FOREWORD BY HOLLIS GOODALL

The MIT Press   Cambridge, Massachusetts   London, England

The MIT Press
Massachusetts Institute of Technology
77 Massachusetts Avenue, Cambridge, MA 02139
mitpress.mit.edu

The MIT Press would like to thank the anonymous peer reviewers who provided comments on drafts of this book. The generous work of academic experts is essential for establishing the authority and quality of our publications. We acknowledge with gratitude the contributions of these otherwise uncredited readers.

This book was set in Futura by the MIT Press. Printed and bound in China.

Library of Congress Cataloging-in-Publication Data is available.

ISBN: 978-0-262-04986-3

10 9 8 7 6 5 4 3 2

EU Authorised Representative: Easy Access System Europe, Mustamäe tee 50, 10621 Tallinn, Estonia | Email: gpsr.requests@easyproject.com

CONTENTS

FOREWORD

Drawing from the ancient and ongoing history of Japanese poetic and pictorial aesthetics, Miya Ando's collection of words and paintings brings the quotidian into the realm of the sublime.

Freezing ephemeral instants of rain quality—some as a metaphor for human emotion—Ando has treated this single phenomenon of rain like a rope, which unfurled reveals innumerable individuated fibers.

Deeply inspired by her Buddhist-priest grandfather and tea-practicing grandmother and mother, Ando on a fundamental level absorbed the pan-Buddhist belief in transitoriness along with the grounding Zen principle of "thusness" (also called "suchness," or the recognition of things just as they are in the moment). In addition, she learned much from the finely shaded aesthetics of *chanoyu* (hot water for tea) and its seasonally differentiated practice. Chanoyu involves using the intellect, for planning the tea ensemble and selecting members of the invited group; the five senses to enjoy the tea, food, flowers, painting and calligraphy, incense, and metal or lacquer utensils in the tearoom; and the movement of the body, as guests enter the tearoom through a low door after passing through a garden and follow a set of prescribed movements during the gathering. Tea is a rare practice in that every element in the tearoom and around it holds aesthetic purpose, and each reflects the other in the overall theme of a single gathering. As the gatherings are never the same, the singular experience had by all in the tearoom is called *ichigo ichie*—one moment, one meeting; or, to treasure the unique moment—a byword for Ando when building the dictionary *Water of the Sky.*

Exemplifying the outlook drawn from Buddhist practice and its association with the transitory and ineffable, Roshi Joan Halifax quotes in this volume from Zen Master Eihei Dōgen (Japan, 1200–1253), who, using the example of water, reveals how perception is based entirely on the observer's situation, and therefore nothing exists as concretely as it appears to any individual.

Contained in Ando's work is an indistinctness that allows for the viewer's mind to expand the context infinitely and to seek their own defining impression. Dōgen asserted further that all time and everything captured within it is momentary; similarly, the fleeting shifts in rain quality are what Ando captures in her word definitions and drawings.[1] The concepts of Zen, including non-duality, transitoriness, and "thusness,"[2] entered Japan in the twelfth century with Dōgen and his predecessor Eisai Myōan (Japan, 1141–1215), who had carried Zen teachings from China in 1191. Aesthetics brought with Zen to Japan at that time were reinterpreted through a native lens, emphasizing the evocative rather than the realistic in studies of nature. Natural, seasonal elements were described as subjects for enjoyment rather than being called upon as metaphors for human emotions, as the court poets from the immediately preceding Heian period (794–1185) did, as noted below. The artistic inclination from this golden age of the court toward suggestion and inference rather than forthright description seeped into art of the Zen style, even as emotive metaphors from nature faded.

A contemporary of Eisai and Dōgen, Kamo no Chōmei (Japan, 1155–1216), initially lived in the ransacked post-civil war imperial capital, Kyoto. After career disappointments and wartime or natural disasters in the capital during his youth, Chōmei became a Buddhist recluse, not yet in the thrall of Zen, but drawn to its emphasis on impermanence (mujō). A quotation from the beginning of Hōjōki, his essays about living in a ten-square-foot hut, reads: "Though the river's current never fails, the water passing, moment by moment, is never the same. Where the current pools, bubbles form on the surface, bursting and disappearing as others rise to replace them, none lasting long. In this world, people and their dwelling places are like that, always changing."[3]

The word *hakanai*, meaning "fleeting," or impermanent, stands as a throughline in Ando's work, whether she depicts rain, clouds, moonlight, or flowers. Describing the contents of her *Water of the Sky* word list, she claims, "The dictionary makes transitoriness observable and examinable. The terms that relate to seasons bind to one's own relationship to time. This type of seasonal experience is shared by everyone and everything."[4] Her choice of materials is alternately reflective or matte, these surfaces playing against one another to alter in varying light.

In the rain series, pure silver and graphite engage the reflective role, while the word defined in the dictionary and the corresponding painted image seize the momentary quality of rain, which is ever-evolving. In terms of her references to poetry and Buddhist writing, Ando says, "Those things that are fragile or short-lived produce the greatest empathy." This statement lies in contrast to much philosophy and poetry of the West and links in Ando's work to the aesthetics of *mono no aware*—"sad beauty"—in poetry, impermanence in Zen, as well as sensitivity to seasonal change, and the patina of use in tea.

Ando's mother, a tea master, introduced Ando to Heian literature at an early age. Ando notes early Heian poetry and writing, from *The Pillow Book* of Sei Shōnagon, *The Tale of Genji*, and particularly *The Kagero Diary*, by Michitsuna no Haha (Japan, c. 935–995), among her favored sources for their pensive beauty and the metaphorical substitution of natural events for human emotions. A quote from Michitsuna no Haha's poetry in the diary gives a clear instance of the type of emotive references found in Ando's dictionary and paintings:

The dew that faded
away is still here on
these sleeves that dry not,
again this morning—the sky
too cannot help drizzling.[5]

Poets of the Heian period, many of them cloistered within palace walls, took note of every detail in their lives. That attention is shown at its utmost in *The Pillow Book* of Sei Shōnagon, who in her notes lists in fine detail everything in her surroundings that pleases or disgusts her. Ando's familiarity with Sei Shōnagon's perception of all phenomena, and Ando's attunement to changes of season passed on to her by her grandmother—who observed in dress and accoutrements the traditional seventy-two seasons in Japan—brought to her attention the finely differentiated aspects of rain.

The drinking of powdered green tea (matcha), encouraged by Eisai in the twelfth century for monks to remain alert during meditation, would over the next three centuries be codified into the practice called chanoyu, which continues to inform much of Japanese culture. The aesthetics that were drawn from Japanese arts of the eleventh and twelfth centuries and filtered through the organizing masters of chanoyu, dominating the arts of the thirteenth through the sixteenth centuries, included concepts of "implication, suggestion, and imperfection."[6] These qualities play a central role in Ando's work and are defined succinctly by Japanese aesthetics historian Yuriko Saito as yojō: emotional aftertaste in poetry that suggests lingering emotions (to which Steve Odin adds mono no aware, "sad beauty," as a subset);[7] wabi: beauty of poverty in tea; sabi: beauty of loneliness in haiku and ink painting; and yūgen: mystery and depth.[8]

Takeno Jōō (Japan, 1500–1552), a merchant, Zen monk, and tea master in the new wabi style of the sixteenth century, emphasized these aesthetics, drawing the practice of tea further toward inclusion of wabi and sabi elements. He was the first to employ the term wabi, in relation to the practice of making tea in a simple hut using local, rather than imported, utensils. (Terms predating wabi, and meaning "chill" and "withered," indicated spareness, timelessness, and solitude.)[9] Takeno, a linked-verse poet himself, drew related aesthetic features from his knowledge of earlier court poetry, particularly that of the poet-courtier Fujiwara no Teika (also known as Fujiwara no Sadaie, Japan, 1162–1241), a contemporary of Zen Master Dōgen. Among Teika's preferred affects within poetry, one finds sabi and yūgen, as well as ushin, "deep feeling." A poem by Teika (c. 1188) in the Senzaishū anthology is reminiscent in its sensitivity to shades of atmosphere that are found in Ando's work:

Fallen rain dripping
From the leaning eaves
So shallow that
Swiftly in pours
The moonlight.[10]

In the second quarter of the fourteenth century, about a hundred years after Teika's death, Yoshida Kenkō (d. 1350) wrote a literary masterpiece in the form of a diary, entitled "Essays in Idleness" (*Tsurezuregusa*, c. 1330–1332). In it, he extolled the patina of age as a feature of the beauty of indistinctness. He explores this theme further in one of the defining phrases of the diary, often quoted by Ando, which reads: "Are we to look at cherry blossoms only in full bloom, the moon only when it is cloudless? To long for the moon while looking at the rain, to lower the blinds and be unaware of the passing of the spring—these are even more deeply moving."[11]

This quotation sums up the preference for the hidden, aged, or incomplete, wrapped up in one of Ando's favorite poetic and pictorial images, *oborozuki*, or the moon partially masked by clouds, the scene evoking for Ando sad beauty, as it also opens the mind to the infinite possibilities of that which lies beyond the mist. In the Buddhist mind, the moon envisioned as obscured and then emerging from mists or clouds brings one closer to the reality of things as they exist: ever-changing.

The aesthetics described above are not the only ones that appear in the arts of Japan, as brilliant screen paintings on gold-leafed paper grounds defined the masterpieces of Japanese art from the late sixteenth century, and brightly ornamental works were collected by samurai and commoners up to the modern era, once the fashion for wabi tea slid from preeminence later in the seventeenth century.[12] The aesthetics of wabi, sabi, yojō with mono no aware, and yūgen remain at the root of Miya Ando's work and in much that is beloved about minimalist expression in the arts of Japan. Because she searches for the essential character of rain and what gives each rain type its individual identity such that it could be named, Ando's process suggests that she has lost her ego to the rain, a step on the path toward Buddhist enlightenment. Rain is an occluding element that dims from our view surroundings, voices, light, and the moon, and so Ando's search for specific patterns of rain hints at clarity of thought or perception in what is otherwise a shadowy realm. A contemporary visual artist who can call on both words and thought patterns from the ancient past that explicate a phenomenon we might see outside our window is both unique and moving. With the *Water of the Sky* dictionary and its paintings, Ando creates an experience of the timeless present.

HOLLIS GOODALL

PREFACE

複雑なことを学ぶよりまず、風や雨、雪や月から送られてくる恋文を読むことを
覚えましょう（一休宗純）

Let us learn to read the love letters sent from the wind, rain, snow, and moon,
rather than studying complicated things, such as sutras.

—Ikkyu Sōjun

The heaviest rain I have ever experienced occurred when I was an adolescent living deep in a redwood forest in the Santa Cruz mountains. As this torrential rain pounded thunderously on our roof, the accumulated force of tiny, innumerable raindrops swayed a redwood so powerfully that the majestic, 200-foot tree came crashing down, its impact reverberating for miles—a sound I will never forget. Redwoods are the tallest trees on earth and can live for hundreds, sometimes even thousands, of years, but Zen'nyo Ryūō—the deity of rain—prevailed that day.

I'm a Japanese and American artist and was raised between two cultures, languages, and geographies. My art is rooted in the dialectic coexistence of Eastern and Western cultures through the lens of natural phenomena. For the past twenty years, I've dedicated my art practice and research to observing, chronicling, and visually articulating manifestations of *mono no aware* in nature. Translating loosely to "an empathy for the pathos of transitory things," mono no aware is a Japanese philosophical concept, arising from Buddhist metaphysics, that recognizes the fundamental nature of reality as impermanent. The natural world is the one expression of this simple but profound way of thinking, offering apt metaphors for the endlessly changing and impossible-to-contain human condition. Take, for example, the Zengo (Zen idiom) "Ame narazu shite hana nao otsu" (Even if there is no rain, petals will fall) (Drawing 1559). This austere adage deftly demonstrates the fleeting or *hakanai* (儚い) quality of circumstances passing ("petals will fall"), despite not having a clear force of change ("even if there is no rain"). It reminds us to be aware of the present and calls attention to the precarious beauty of each fleeting moment.

In this lineage, the subjects of my creative focus are particularly evanescent natural phenomena: *Shinrabansho* 森羅万象 ("the all-covering forest," i.e., all things in nature that exist under heaven)—clouds, rain, shooting stars, the waxing and waning of the moon, as well as their etymological and textual expressions, that is, how these phenomena have been described across time. Japanese poetry and literary traditions offer rich histories of writers describing their surroundings with encyclopedic specificity and hyperlocal awareness of the changing micro-seasons.[1] I'm likewise interested in articulating the minute nuances of my surroundings. My engagement with these subjects is meditatively serial, accumulating indexically into vast visual compendiums that evoke thousands of Japanese-language idioms through my materials.

My entire canon of visual work is in pursuit of this conceptual paradox: to capture something fleeting, not as a way to contain it but as an ode to the universality of impermanence, up to and including our own mortality.

Through a collection of 2,000 Japanese words and their English interpretations, *Water of the Sky* describes the breadth and diversity of rain's many expressions: when it falls, how it falls, and how its observer might be transformed physically or emotionally by its presence. The words range from the prosaic to esoteric, extending from the meteorological (Mukaame [Very Fine Rain That Falls in Spring]) to the mystical (Bunryūu [Rain That Splits a Dragon's Body in Half]), and from the minute (Kisame [Raindrops That Fall Off the Leaves and Branches of Trees]) to the vast (Takuu [Blessed Rain That Quenches All Things in the Universe]).[2] My visual interpretations of these terms are not so much illustrations as evocations, attempts to embody or imagine that particular rain's precise and essential quality.

Water of the Sky presents 100 of these 2,000 drawings of rain,[3] limited in number due to their sheer volume and the inevitable spatial constraints of a print book. The drawings are accompanied by the full index of 2,000 Japanese words and their approximate English equivalents, titled numerically by the order in which the drawings were made and ultimately categorized alphabetically using a hybrid Japanese-English alphabetization rubric that categorizes Japanese phonemes by the English A-B-C, a perhaps humorously derivative method that's a consequence of my bilingual mind.[4]

Water of the Sky acquires its title not from the typical vernacular word for sky, *sora* (空), but via *ama* (天), more often used to describe the heavenly realm.[5] "Ama no mizu" is the ancient way to say rain and translates to "water of the heavens" or "water that belongs to the sky." The 2,000 words gathered and presented in this book are reflective of my scholarly interests and lived experience: I grew up (in part) in my family's Buddhist temple in Okayama, Japan, and therefore am partial to words pertaining to Buddhist rain sutras, rain ceities, and Amagoi (prayers for rain). As an admirer of poetic anthologies, I have included observations of seasonal change (*kigo*) and have prioritized observations from Heian-period women poets frequently sidelined in mainstream literary culture. Another influence in my criteria for word selection is my mother, a teacher of the Urasenke school of tea ceremony. Her guidance and philosophy drew me to the many Zengo contained in this book, added not as terms describing particular types of rain but for their ability to communicate rain as a metaphor for the nature of existence. Lastly, recognizing the unique perspectives of lesser-known dialects, having grown up speaking the Okayama (Chūgoku) dialect as my first language, I've also included several regionally specific words from throughout Japan, vernacular that tends to be particularly whimsical and reflects micro-localized perceptions. The terms span a timeline of over 1,000 years, from as far back as the *Nihon Shoki*, a text compiled in 720 CE, and the *Man'yōshū*, or *Collection of Ten Thousand Leaves*,[6] to as recent as the term Kuroi Ame (A Black Rain That Fell in Hiroshima Immediately after the Atomic Bomb). While I present this as a bilingual dictionary with an equally essential visual component, I recognize the subjectivity of language interpretation as informed by these influences.

My bicultural vantage point has been vital in presenting this work, as this compendium abounds in lacunae. Lexical voids, or gaps left when a word exists in one language but not another, lacunae can be a way of examining differing cultural value systems. Not only is a word missing, but so too is the entire concept it represents. When something is valued, it is given a name. The Japanese language demonstrates such acute obser-vations, attunement, and reverence for nature wherein, for example, there exist seventeen unique words to express a particular type of soft, quiet, elegant rain.[7] Many of the terms I present in this book are anachronistic, no longer part of the Japanese lexicon, and this has presented a particularly challenging aspect of translation. These words, like the fleeting rains they describe, are equally prone to recede into the depths of time. I've made it my prerogative to steward them, to deliver them to the next generation of scholars and readers, so that, in the changing tides of culture that dem-onstrate our shifting values, their wisdom is not lost. Beyond making these arcane words available to contemporary Japanese and English speakers alike, I am also aware of how this work functions as a time capsule or an extinction diary for the types of rains themselves as we face changes in our climate and weather patterns. I hope this text provides a point of entry to readers with interests ranging from linguistics and lexicology, language his-tory and preservation, Japanese literature, biracial artistic enterprise and multiculturalism, science communication, nature, conceptual art and serial structures, cultural anthropology, contemporary art, and drawing.

Ultimately, I am an artist and this is an artist book. To me, art has always functioned as the materialization of thought, which is why the visual compo-nent of this compendium is essential. The drawings make the words visible, anecdotal, symbolic, and referential. They are a gateway to those without language, and a tool for approaching lacunae when it's one's own lan-guage that holds the gap. Rain is a transitory connective thread between the heavenly and earthly planes, here and then immediately gone, a per-fect example of mono no aware and in some ways quite difficult to repre-sent. The drawings presented here are made using traditional natural indigo dye. Used globally for over 5,000 years, it was introduced to Japan during the Heian period (794–1185 CE). Besides the obvious association of blue with water and the sky, indigo makes an apt material also for the way that it functions as a type of clock: the longer a material is submerged in this dye, the deeper the hue. Conceptually, therefore, it is a material that both records and communicates time's passage.[8] In addition to indigo dye, I have employed pencil and micronized pure silver to execute the drawings.

Finally, this project is an ode to the speechless quality of rain, perhaps ironic given the project's emphasis on language. Throughout the three years during the Covid pandemic when I wrote this dictionary, I frequently returned to my childhood memory of learning the phrase, "When it rains while the sun is shining, foxes are having a wedding ceremony." I was shocked that there existed a word so specific for that type of rain.[9] That wonder and ardent respect for nature have remained with me all this time, inspiring me to be steadfast in my dedication to this project even when it became arduous. In such moments, I found quietude by "listening carefully to hear what the rain is saying"[10] and perceiving the fleeting beauty of our brief time on earth. Perhaps you too can "listen to the sound of rain as it seeps into your heart and soul"[11] and be calmed by our precious, perfect impermanence.

MIYA 美夜 ANDO

PERCEIVING RAIN

a wanderer,
  let that be my name—
  the first winter rain.
  Matsuo Bashō

In Zen, an *unsui* (clouds/water) is one who drifts freely like a cloud and flows unimpeded like water. As the great eighth-century Zen Master Shitou Xiqian uttered, "The vast sky does not hinder the white clouds from flying." And so it is with these images of clouds, rain, and space. One drifts freely through the artist's years of rain-soaked moments and experiences infinite indigo hues that express the mysterious expressions of rain, rain upon rain, the utter transitoriness of rain, the silence and sound of rain, this evanescent gift of heaven.

The Christian monk Thomas Merton too met rain, again and again, in the thick forest surrounding his isolated hermitage in the dark and soaking Kentucky woods. He wrote:

> What a thing it is to sit absolutely alone, in the forest, at night, cherished by this wonderful, unintelligible, perfectly innocent speech, the most comforting speech in the world, the talk that rain makes by itself all over the ridges, and the talk of the watercourses everywhere in the hollows!
>
> Nobody started it, nobody is going to stop it. It will talk as long as it wants, this rain. As long as it talks I am going to listen.[1]

Miya Ando's images speak the language of rain, rain expressing the alchemy of change, of creation, preservation, and dissolution. She knows that without rain, there is no life, no earth, no birth, no death, no coming forth, and no decay. Life and death are born out of rain, are born of rain, rain that falls on mountain summits, rain that joins creeks and rivers, rain that disappears into the ink of the night ocean, rain that captures prismatic light, rain that never touches the earth, rain on blossom petals and bent stalks of green rice shoots, rain that washes away salty tears and cleans muddy farm-worked hands. Rain, though lonely, is accompanied by others like itself, and also by clouds, and by wind, thunder, and lightning.

And, as we see from Miya Ando's images, rain depends on the perceiver. In Zen Master Dōgen's "Mountains and Waters Sutra," he writes:

> Some beings see water as wondrous blossoms, but they do not use blossoms as water. Hungry ghosts see water as raging fire or pus and blood. Dragons see water as a palace or a pavilion. Some beings see water as the seven treasures or a wish-granting jewel. Some beings see water as a forest or a wall. Some see it as the dharma nature of pure liberation, the true human body, or the form of the body and the essence of mind. Human beings see water as water. Water is seen as dead or alive depending on [the seer's] causes and conditions. . . . Thus, the views of all beings are not the same. Question this matter now. The thoroughly actualized realm has one thousand kinds and ten thousand ways. . . . In this way, water is not just earth, water, fire, wind, space, or consciousness. Water is not blue, yellow, red, white, or black. Water is not forms, sounds, smells, tastes, touchables, or the objects of mind. But water as earth, water, fire, wind, and space actualizes itself. . . . Know that even though all things are unbound and not tied to anything, they abide in their own condition. However, when most human beings see water they only see that it flows unceasingly. This is a limited human view; there are actually many kinds of flowing. Water flows on the earth, in the sky, upward, and downward. It flows around a single curve or into many bottomless abysses. When it rises it becomes clouds. When it descends it forms abysses. . . . The path of water is not noticed by water, but is actualized by water. . . . Ordinary fools and mediocre people nowadays think that water is always in rivers or oceans, but this is not so. There are rivers and oceans within water. Thus, even where there is not a river or an ocean, there is water. It is just that when water falls down to the ground, it manifests the characteristics of rivers and oceans. . . . Even in a drop of water innumerable buddha lands appear. . . . Where buddha ancestors reach, water never fails to reach. Thus, buddha ancestors always take up water and make it their body and mind, make it their thought. . . . You who study with buddhas should not be limited to human views when you see water. Go further and study water in the buddha way. Study how you view the water used by buddha ancestors. Study whether there is water or no water in the house of buddha ancestors.[2]

This extraordinary collection of exquisite and ephemeral images is a study of water, a study of rain, a study of the gift of heaven, which gives life to life. You are the perceiver. What do you see?

JOAN HALIFAX
ABBOT, UPAYA ZEN CENTER

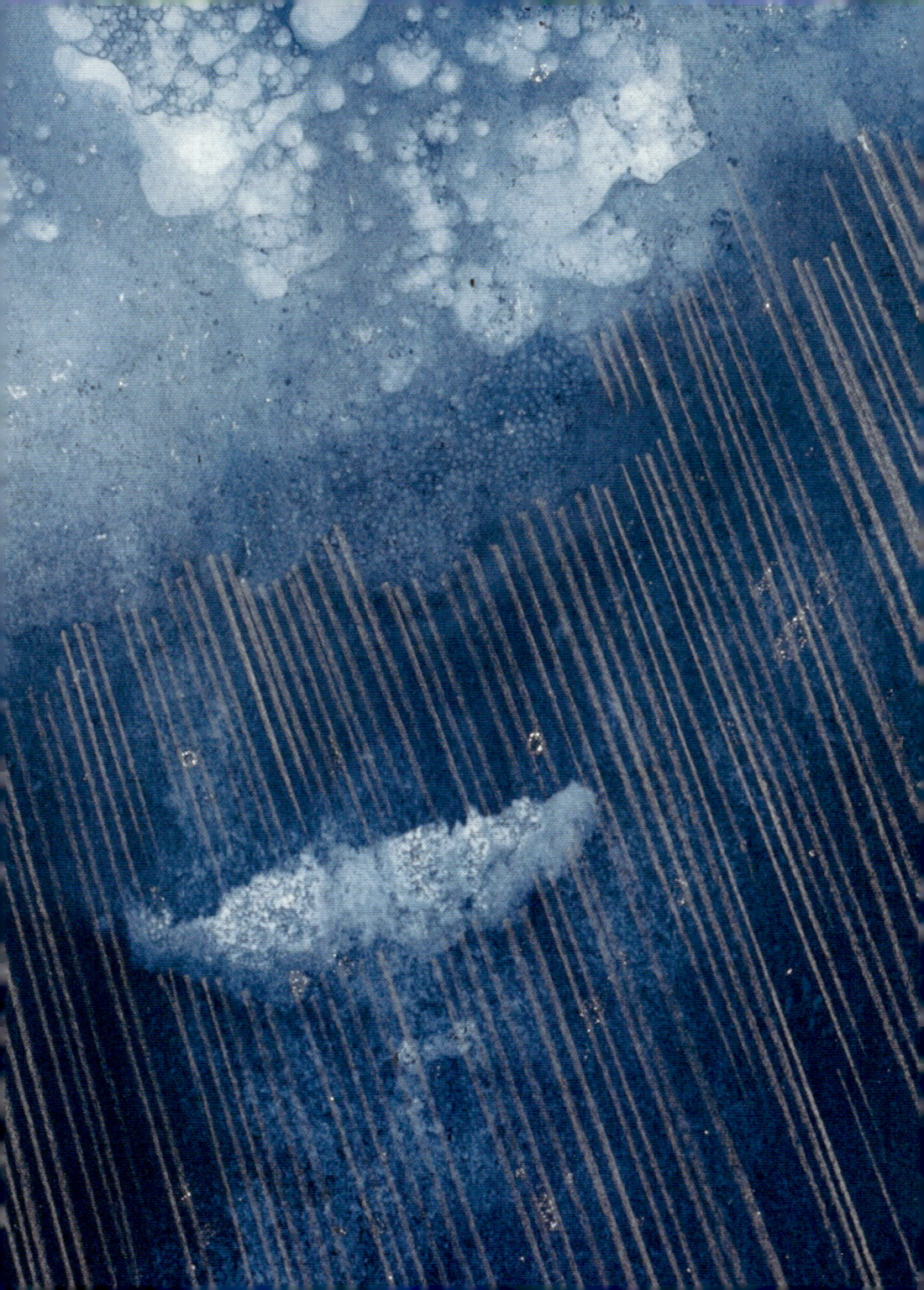

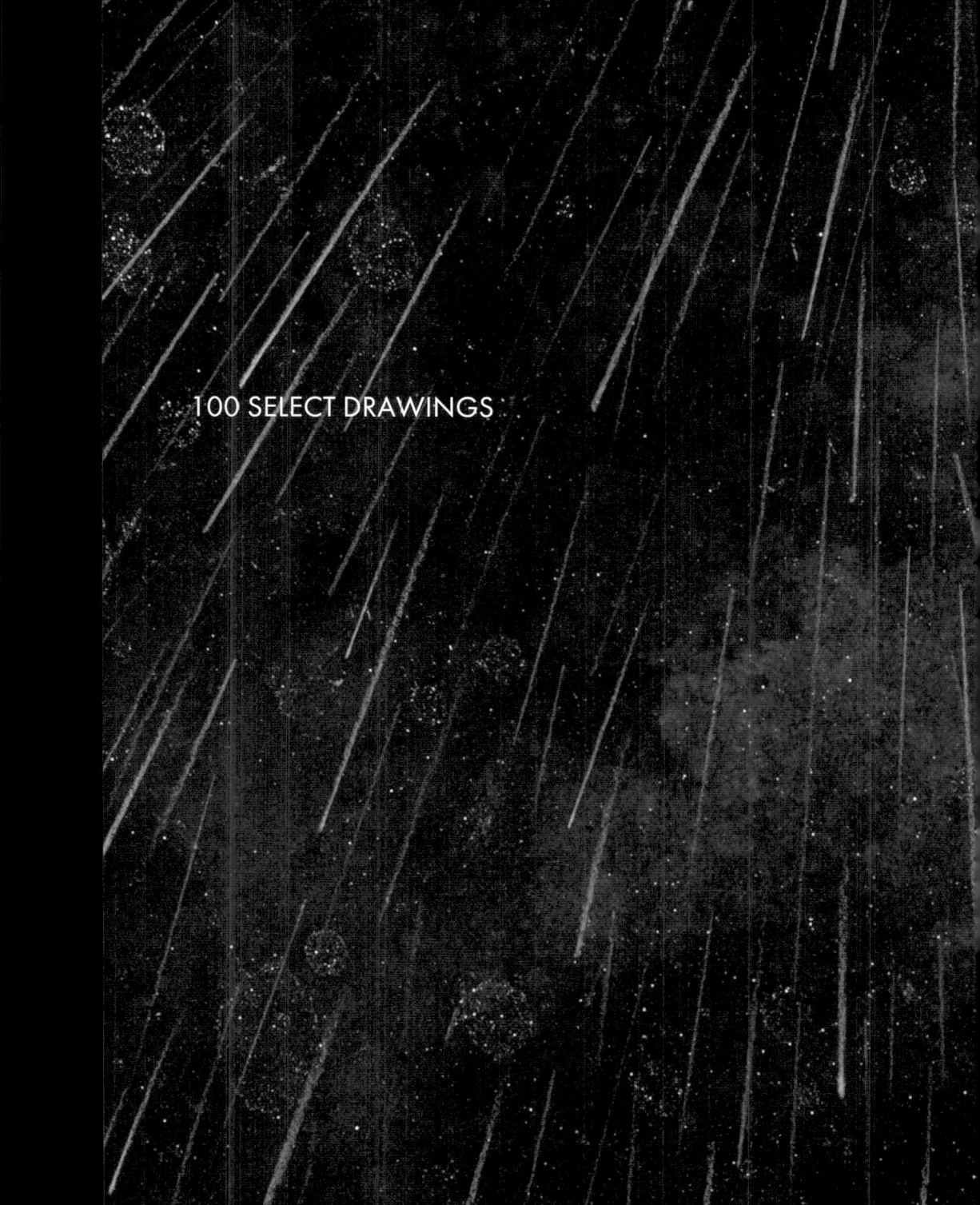

100 SELECT DRAWINGS

0004_Onnaame (Rain That Falls Slowly / Woman's Rain)
Natural indigo dye, micronized pure silver, graphite, Hahnemühle paper
8.5 × 11 in (21.59 × 27.94 cm)
2021

0009_Kyūu (Praying for Rain to Fall)
Natural indigo dye, micronized pure silver, graphite, Hahnemühle paper
11 × 8.5 in (27.94 × 21.59 cm)
2021

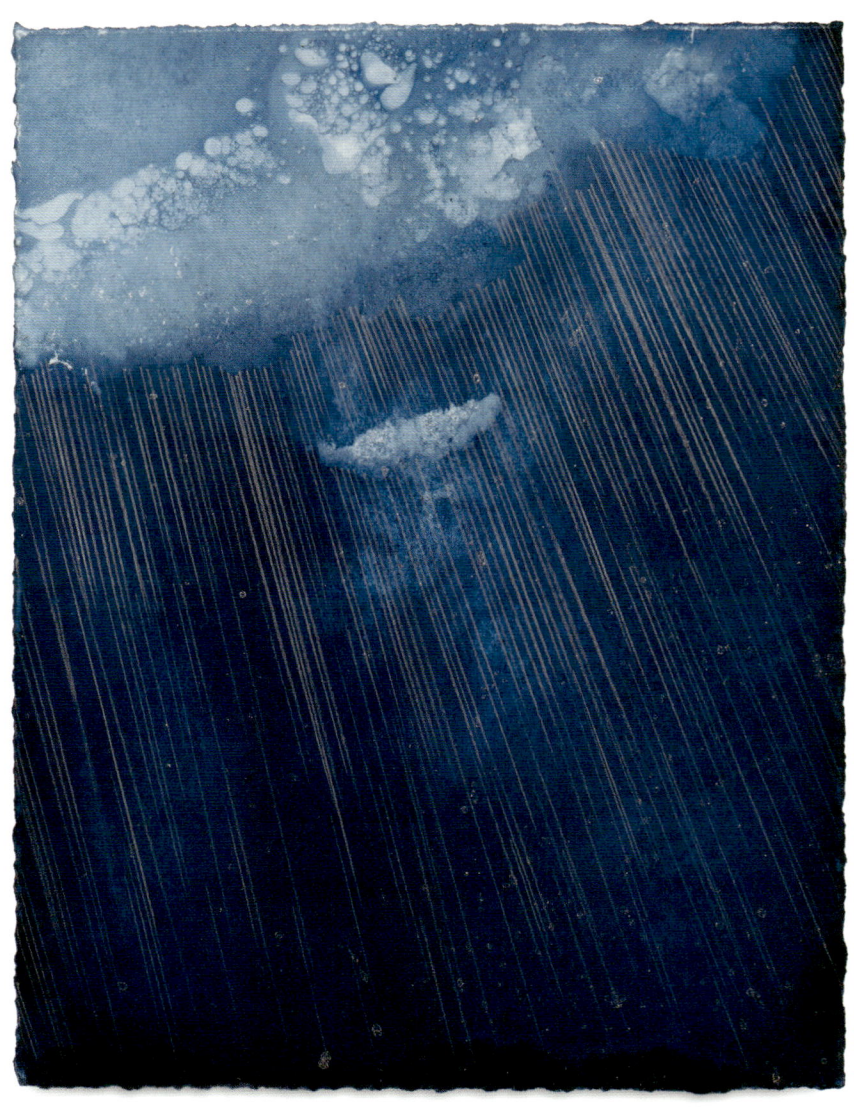

0012_Uryū Ensa (Describes the Appearance of a Fisherman Working in the Rain)
Natural indigo dye, micronized pure silver, graphite, Hahnemühle paper
11 × 8.5 in (27.94 × 21.59 cm)
2021

0021_Keame (Fine Rain That Looks like a Fiber)
Natural indigo dye, micronized pure silver, graphite, Hahnemühle paper
11 × 8.5 in (27.94 × 21.59 cm)
2021

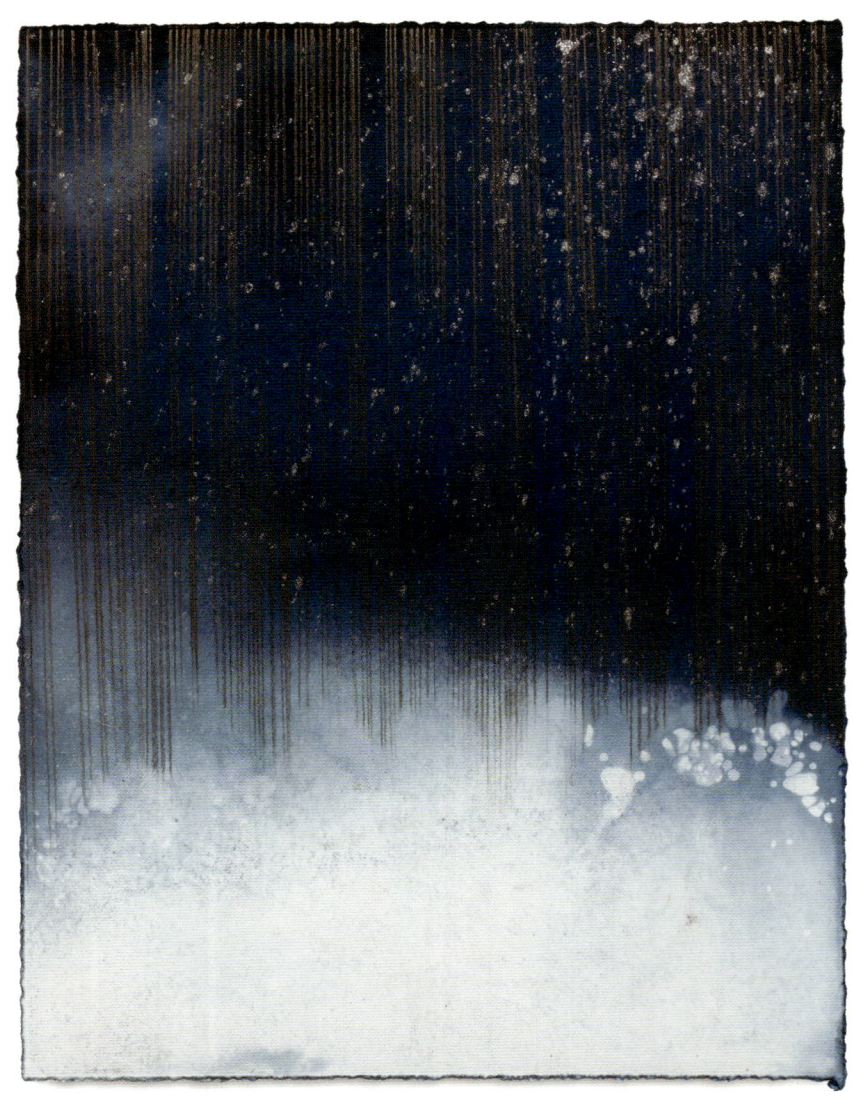

0022_Mukaezuyu (Rain That Lasts Several Days before the Rainy Season
and Ushers in the Rainy Season)
Natural indigo dye, micronized pure silver, graphite, Hahnemühle paper
11 × 8.5 in (27.94 × 21.59 cm)
2021

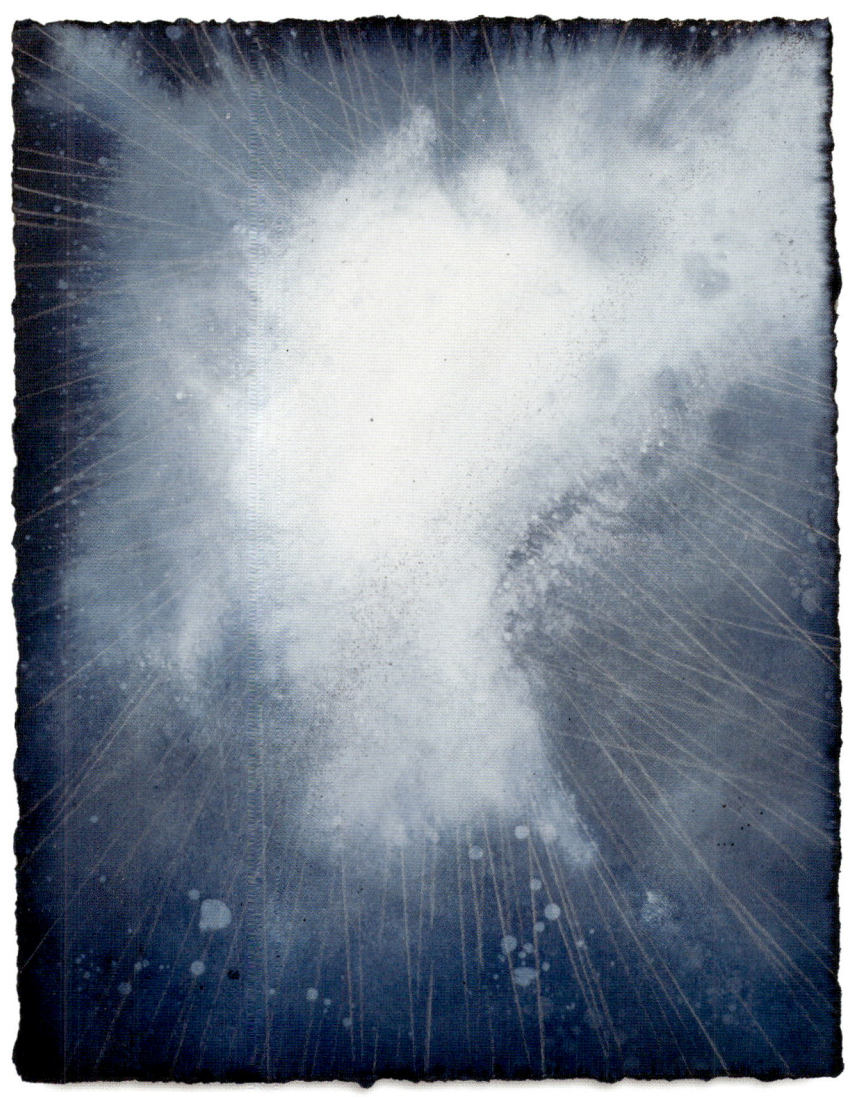

0025_Shūu (A Sudden Shower / Cloudburst)
Natural indigo dye, micronized pure silver, graphite, Hahnemühle paper
11 × 8.5 in (27.94 × 21.59 cm)
2021

0026_Amatsutsumi (Being Trapped by Rain and Unable to Go Outside)
Natural indigo dye, micronized pure silver, graphite, Hahnemühle paper
11 × 8.5 in (27.94 × 21.59 cm)
2021

0028_Koburi (It Rains a Little Bit and Is Weak)
Natural indigo dye, micronized pure silver, graphite, Hahnemühle paper
11 × 8.5 in (27.94 × 21.59 cm)
2021

0036_Fuyushigure (Light Winter Drizzle)
Natural indigo dye, micronized pure silver, graphite, Hahnemühle paper
11 × 8.5 in (27.94 × 21.59 cm)
2021

0065_Ame no Hana (Cherry Blossoms in the Rain / Flowers in the Rain)
Natural indigo dye, micronized pure silver, graphite, Hahnemühle paper
11 × 8.5 in (27.94 × 21.59 cm)
2021

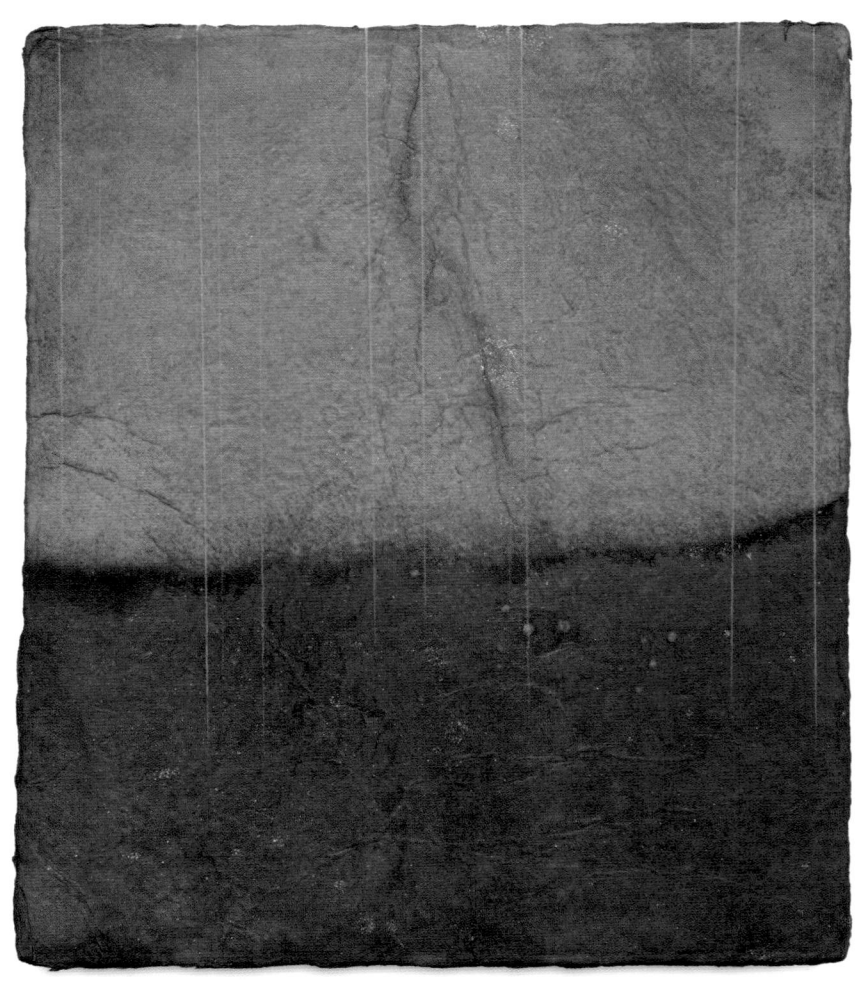

0094_Matsukaze no Shigure (Sound of Drizzle in the Pines)
Natural indigo dye, micronized pure silver, graphite, Kozo paper
10.5 × 9.5 in (26.67 × 24.13 cm)
2021

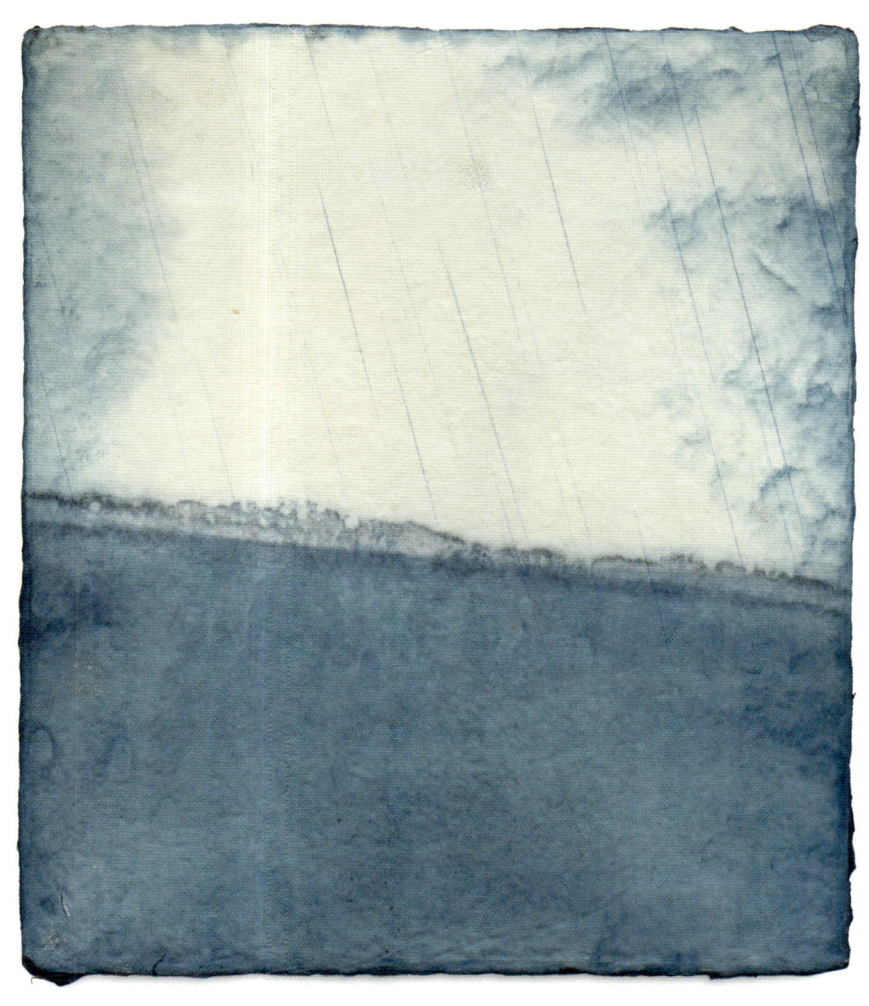

0106_Yamameguri (Rain n the Mountains)
Natural indigo dye, micronized pure silver, graphite, Kozo paper
10.5 × 9.5 in (26.67 × 24 13 cm)
2021

0111_Shūame (Rain Falling from Clouds)
Natural indigo dye, micronized pure silver, graphite, Kozo paper
10.5 × 9.5 in (26.67 × 24.13 cm)
2021

0125_Fujinoyama Arai (Washing of the Mountain Rain / Strong Rain in Autumn /
A Rain That Cleans Mount Fuji)
Natural indigo dye, micronized pure silver, graphite, Hahnemühle paper
11 × 8.5 in (27.94 × 21.59 cm)
2021

0126_Kitsune no Yomeiri (Rain That Falls Even Though the Sun Is Shining /
The Day That Foxes Have Their Wedding Ceremony)
Natural indigo dye, micronized pure silver, graphite, Kozo paper
10.5 × 9.5 in (26.67 × 24.13 cm)
2021

0128_Fuyuame (Winter Rain, like Sleet)
Natural indigo dye, micronized pure silver, graphite, Hahnemühle paper
11 × 8.5 in (27.94 × 21.59 cm)
2021

0129_Tanabataame (Rain That Falls during the Tanabata Star Festival Night of July 7)
Natural indigo dye, micronized pure silver, graphite, Hahnemühle paper
11 × 8.5 in (27.94 × 21.59 cm)
2021

0131_Sowofuru (Elegant and Quiet Rain)
Natural indigo dye, micronized pure silver, graphite, Hahnemühle paper
11 × 8.5 in (27.94 × 21.59 cm)
2021

0140_Fuji no Ame (Wisteria Rain / Rain That Falls When Wisteria Flowers Are in Bloom)
Natural indigo dye, micronized pure silver, graphite, Hahnemühle paper
11 × 8.5 in (27.94 × 21.59 cm)
2021

0168_Sayoshigure (A Weak and Light Rain Shower in the Night)
Natural indigo dye, micronized pure silver, graphite, Kozo paper
9.5 × 10.5 in (24.13 × 26.67 cm)
2021

0172_Amaochi Byōshi (Rhythm That Mimics the Raindrops Falling from the Roof,
Used When Learning Shamisen)
Natural indigo dye, micronized pure silver, graphite, Kozo paper
9.5 × 10.5 in (24.13 × 26.67 cm)
2021

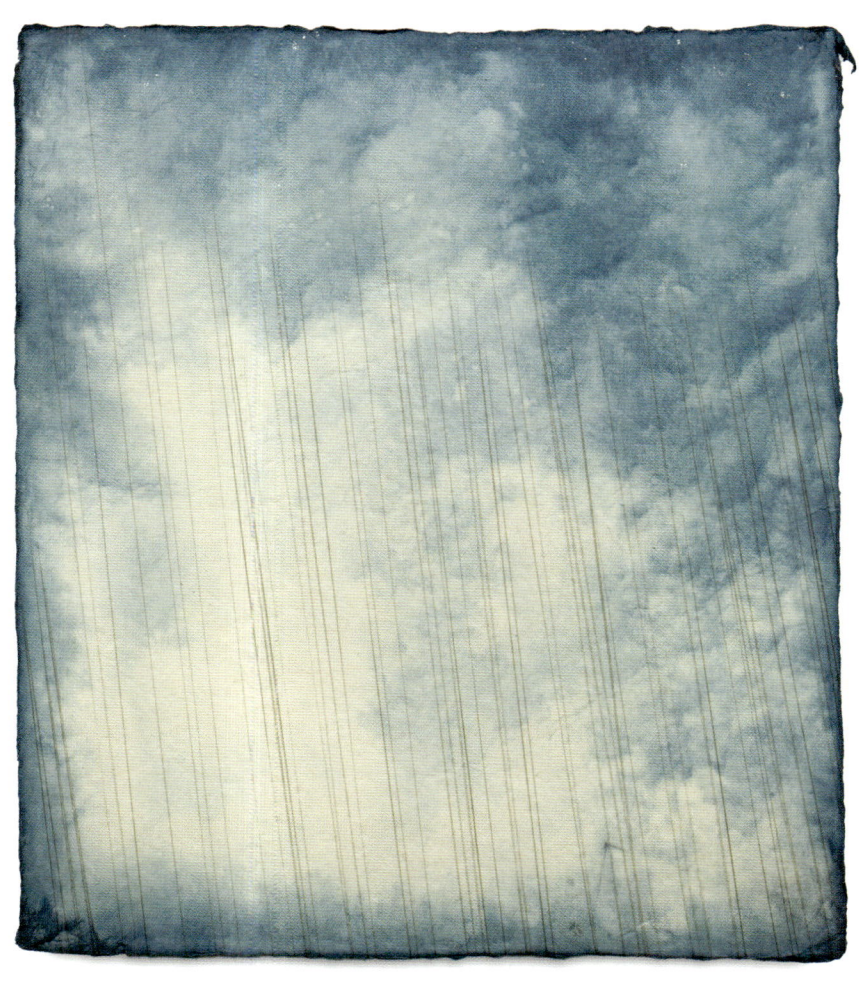

0180_Ginga Tōsha (Very Heavy Rain or a Large Waterfall / A Galaxy Descending to Earth)
Natural indigo dye, micronized pure silver, graphite, Kozo paper
10.5 × 9.5 in (26.67 × 24.13 cm)
2021

0189_Yau (Night Rain)
Natural indigo dye, micronized pure silver, graphite, Hahnemühle paper
11 × 8.5 in (27.94 × 21.59 cm)
2021

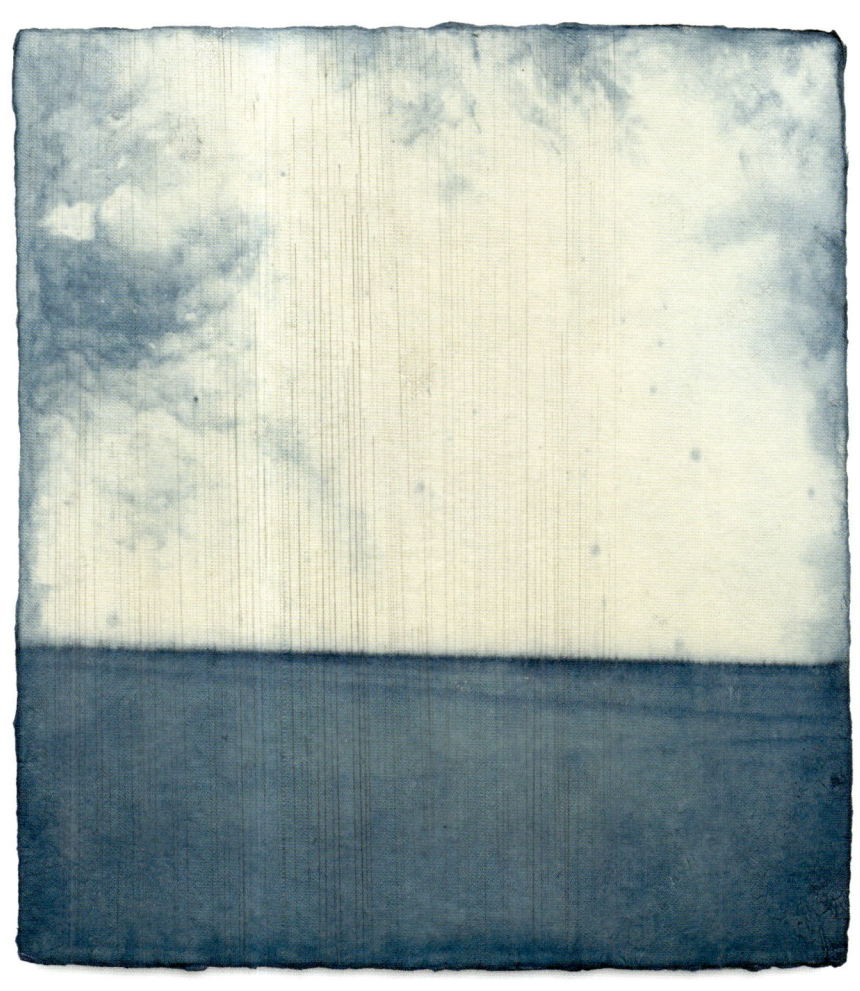

0201_Koame (A Tiny Little Light Rain)
Natural indigo dye, micronized pure silver, graphite, Kozo paper
10.5 × 9.5 in (26.67 × 24.13 cm)
2021

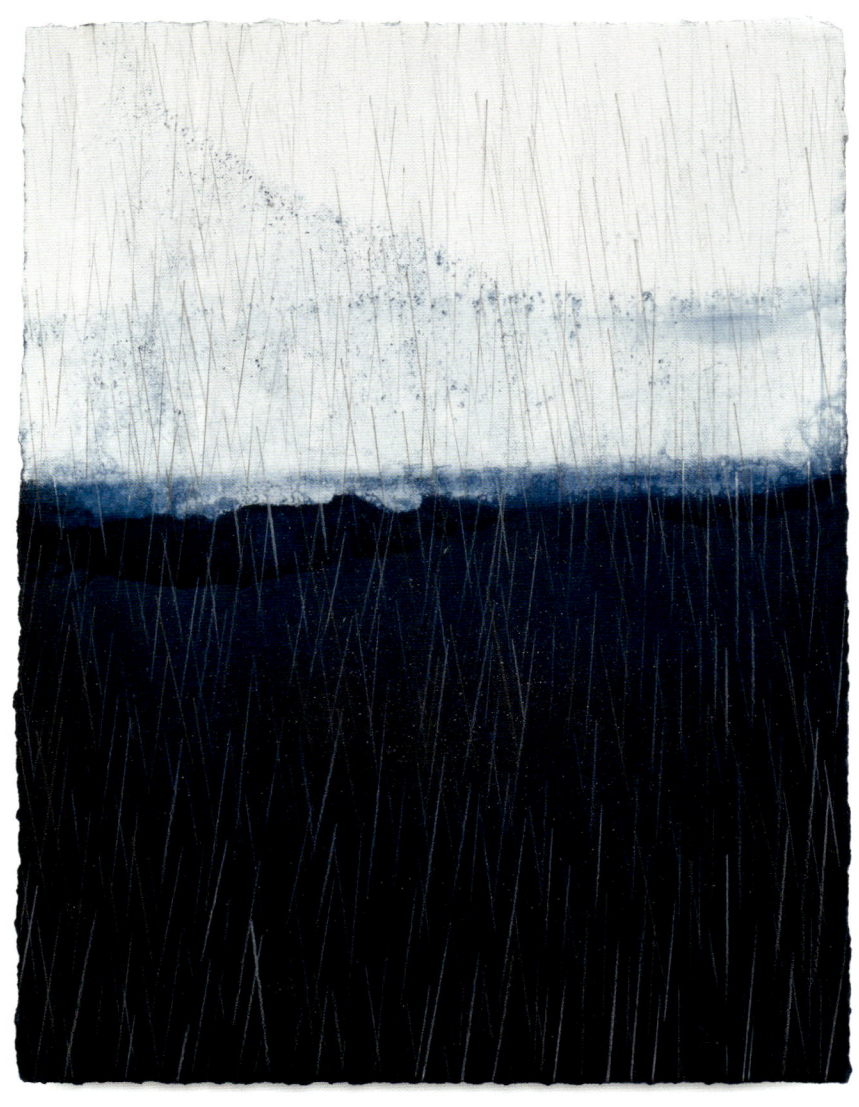

0215_Hakushūu (Heavy Rain That Falls Intermittently in Autumn)
Natural indigo dye, micronized pure silver, graphite, Hahnemühle paper
11 × 8.5 in (27.94 × 21.59 cm)
2021

0222_Tsukishigure (Rain in the Moonlight)
Natural indigo dye, micronized pure silver, graphite, Hahnemühle paper
11 × 8.5 in (27.94 × 21.59 cm)
2021

0227_Aiaigasa (Sharing an Umbrella with Someone / Romantic Rain)
Natural indigo dye, micronized pure silver, graphite, Hahnemühle paper
11 × 8.5 in (27.94 × 21.59 cm)
2021

0252_Potsuripotsuri (Rain Falling in Drops)
Natural indigo dye, micronized pure silver, graphite, Hahnemühle paper
11 × 8.5 in (27.94 × 21.59 cm)
2021

0255_Yukikeshi no Ame (Rain That Erases Snow)
Natural indigo dye, micronized pure silver, graphite, Hahnemühle paper
11 × 8.5 in (27.94 × 21.59 cm)
2021

0256_Jiu (Late Fall or Ecrly Winter Rain Shower / Welcome Rain and Beneficial Rain)
Natural indigo dye, micrcnized pure silver, graphite, Hahnemühle paper
11 × 8.5 in (27.94 × 21.59 cm)
2021

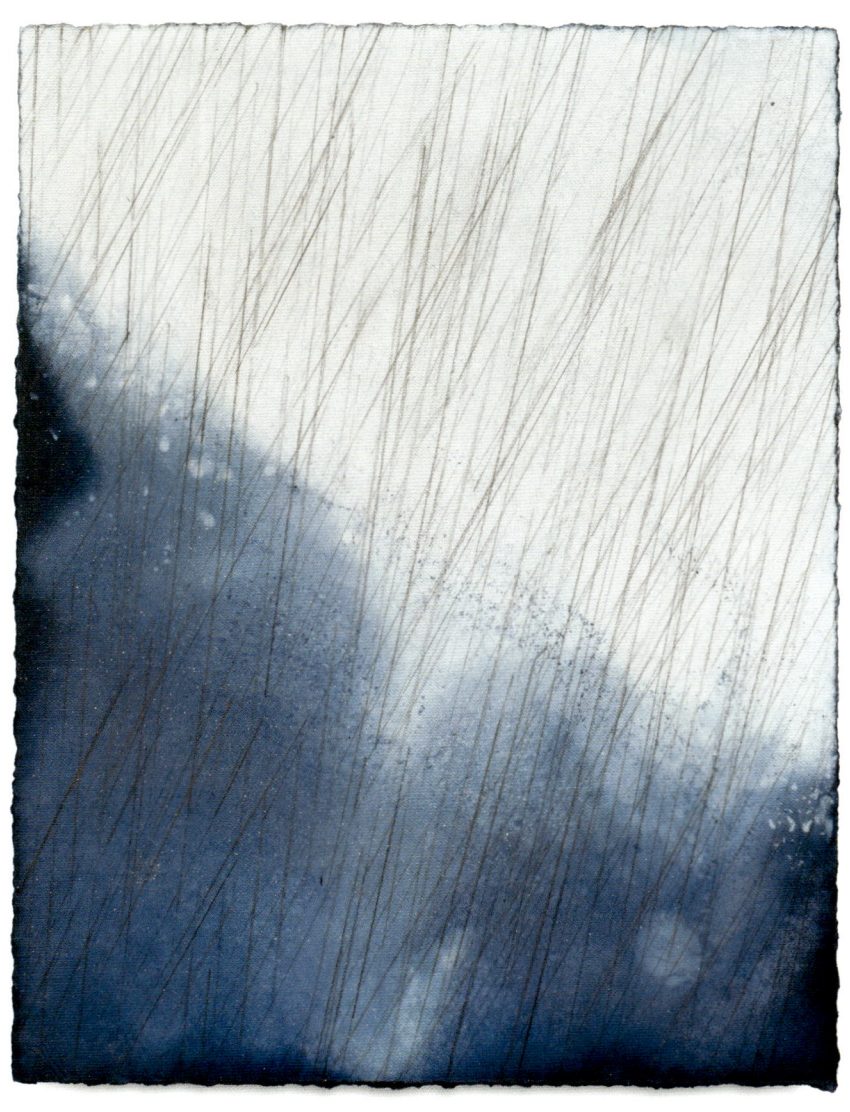

0257_Murasame (Passing Rain / It Pours Heavily and Stops Immediately)
Natural indigo dye, micronized pure silver, graphite, Hahnemühle paper
11 × 8.5 in (27.94 × 21.59 cm)
2021

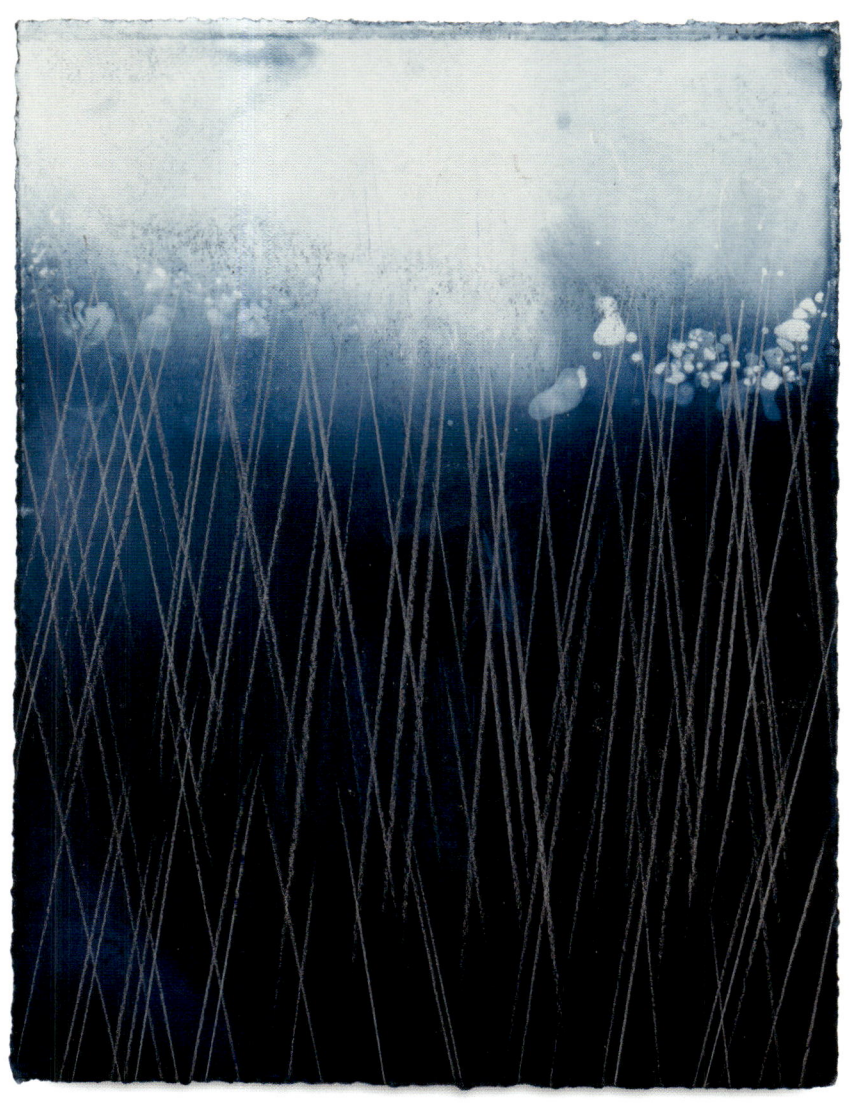

0259_Manjyōfū'u (An Entire Town Is Hit with Rain and Wind)
Natural indigo dye, micronized pure silver, graphite, Hahnemühle paper
11 × 8.5 in (27.94 × 21.59 cm)
2021

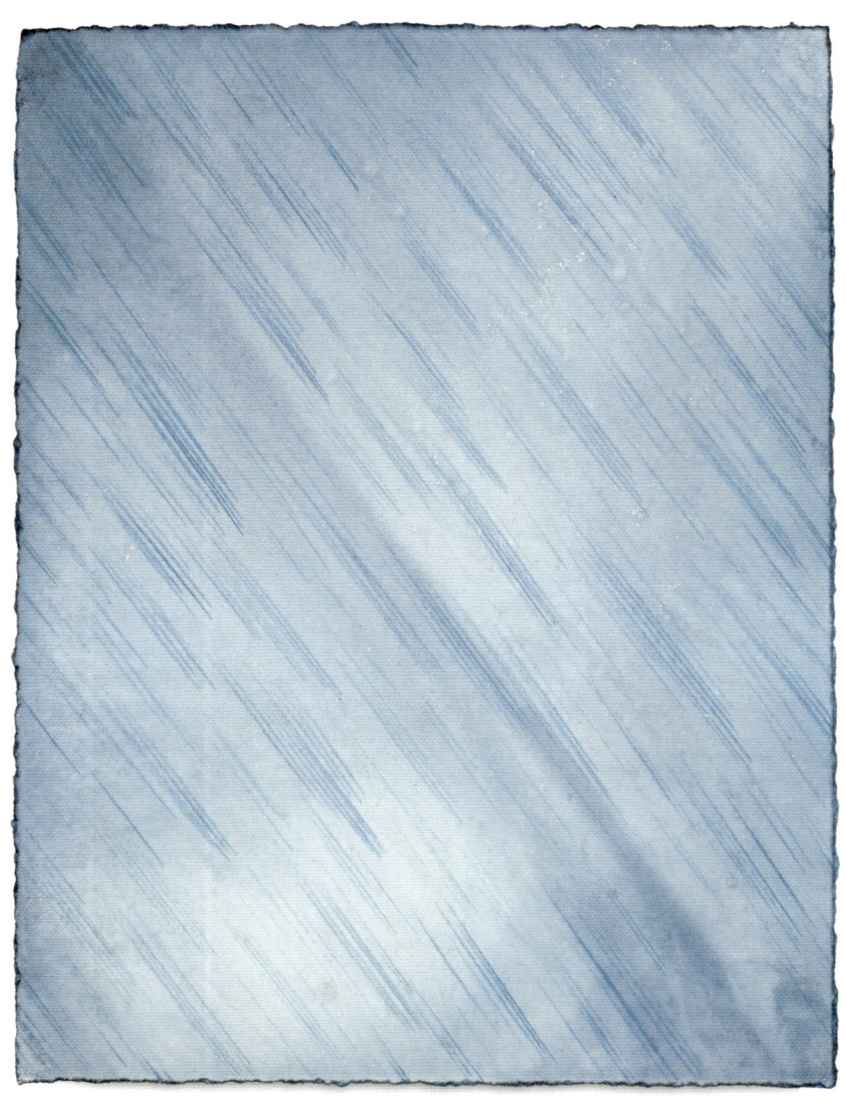

0260_Aoshigure (Blue Rain)
Natural indigo dye, micronized pure silver, graphite, Hahnemühle paper
11 × 8.5 in (27.94 × 21.59 cm)
2021

0261_Ame ya Arare to (Rain and Hail)
Natural indigo dye, micronized pure silver, graphite, Hahnemühle paper
11 × 8.5 in (27.94 × 21.59 cm)
2021

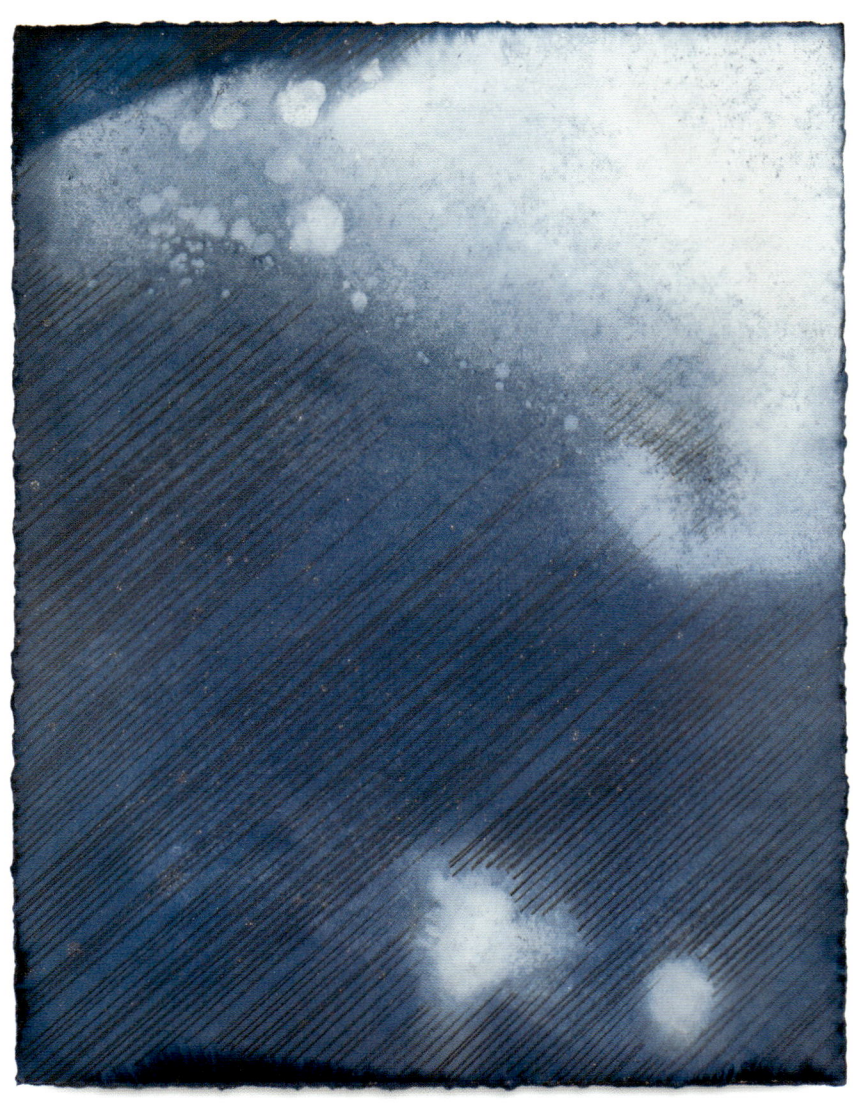

0262_Akitsuiri (Autumn Rainy Season)
Natural indigo dye, micronized pure silver, graphite, Hahnemühle paper
11 × 8.5 in (27.94 × 21.59 cm)
2021

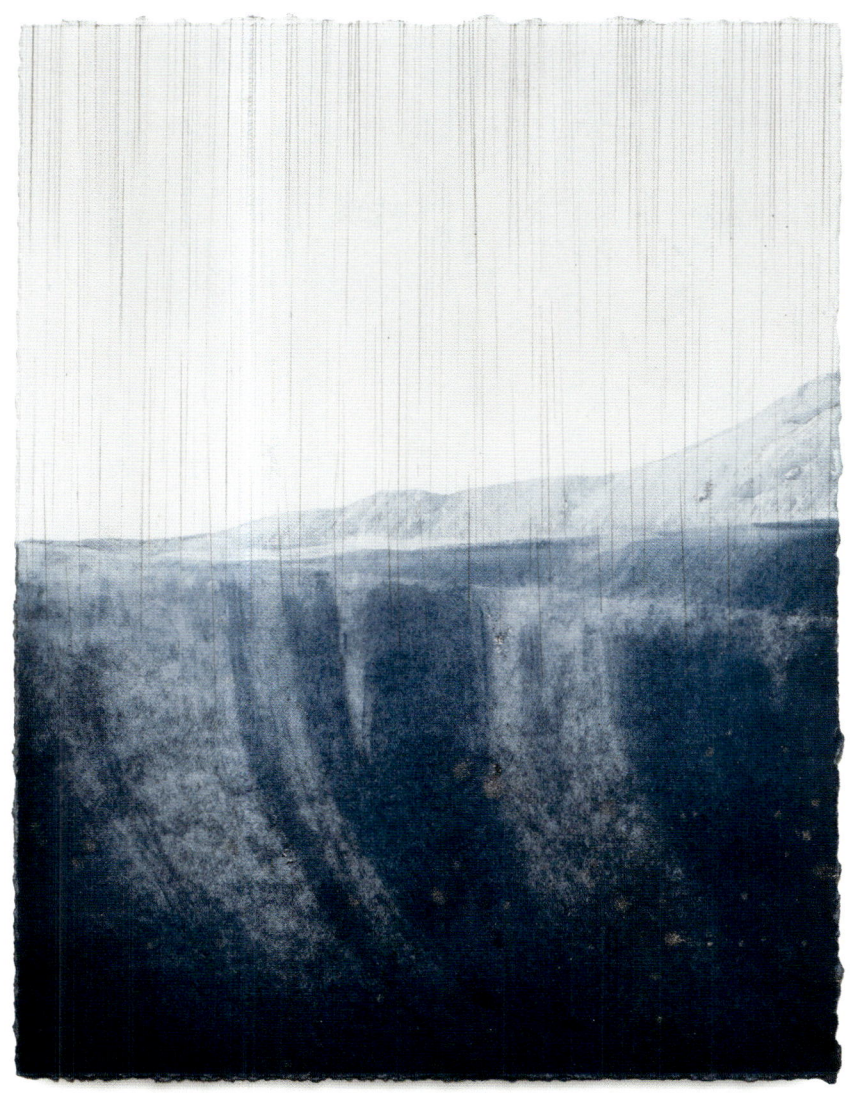

0264_Yoka no Ame (Rain Falling on the Cherry Blossoms High atop the Cold Mountains That Do Not Bloom until Early Summer)
Natural indigo dye, micronized pure silver, graphite, Hahnemühle paper
11 × 8.5 in (27.94 × 21.59 cm)
2021

0266_Akishimeri (A Long Spell of Rain in Autumn)
Natural indigo dye, micronized pure silver, graphite, Hahnemühle paper
11 × 8.5 in (27.94 × 21.59 cm)
2021

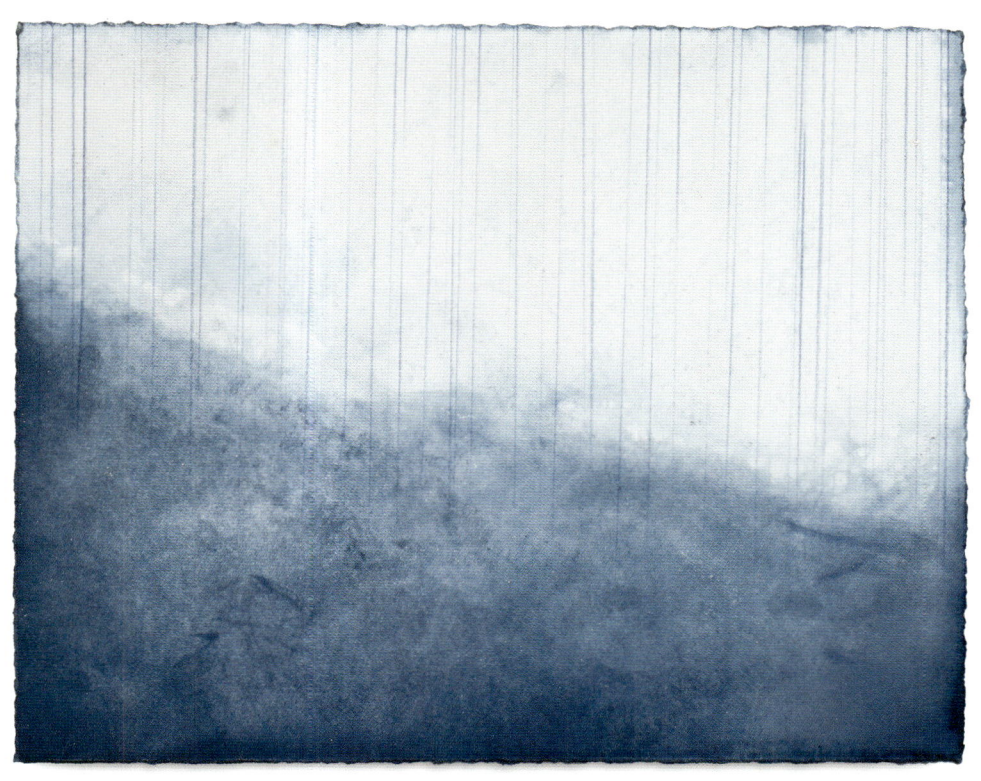

0286_Aobaame (Rain That Falls in Early Summer on Green Leaves / Blue Leaf Rain)
Natural indigo dye, micronized pure silver, graphite, Hahnemühle paper
8.5 × 11 in (21.59 × 27.94 cm)
2021

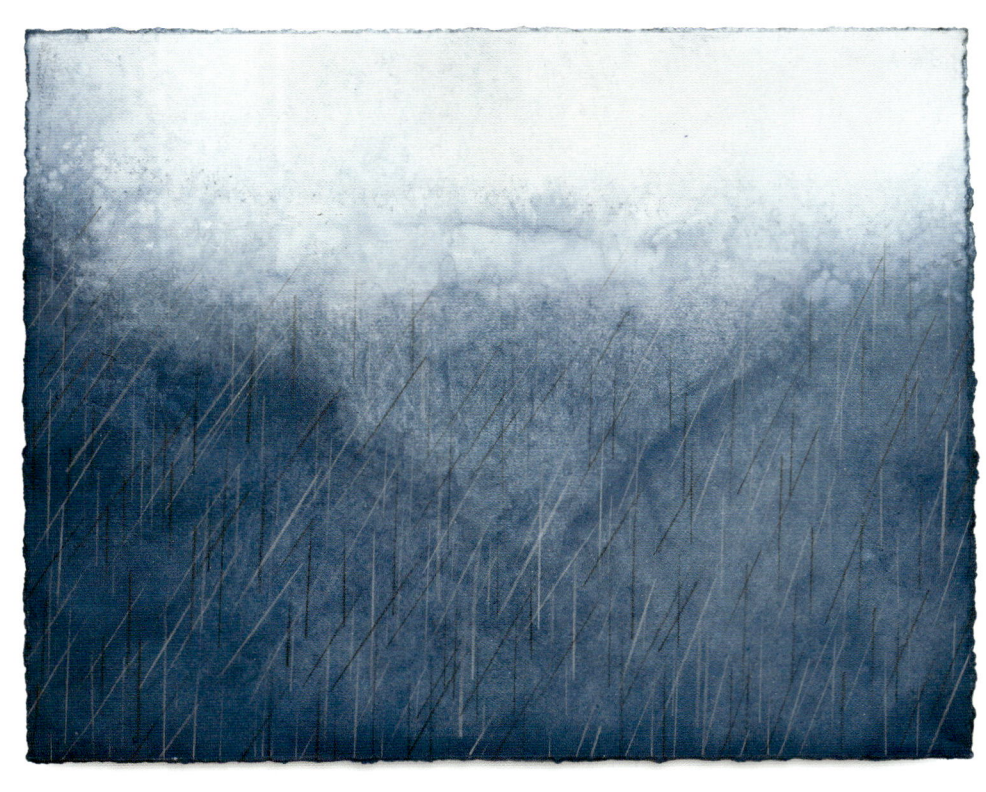

0296_Ōrui (Blessed Rain—Kagawa)
Natural indigo dye, micronized pure silver, graphite, Hahnemühle paper
8.5 × 11 in (21.59 × 27.94 cm)
2021

0298_Shinotsukuame (Intense Rain That Falls Heavily, Is Very Fine and Strong
like the Bamboo Grove at Shinotake)
Natural indigo dye, micronized pure silver, graphite, Hahnemühle paper
8.5 × 11 in (21.59 × 27.94 cm)
2021

0331_Aki no Nagaame (Long Rain of Autumn)
Natural indigo dye, micronized pure silver, graphite, Hahnemühle paper
11 × 8.5 in (27.94 × 21.59 cm)
2021

0332_Amagakeru (Soaring Rain / Gods, Souls, Birds Fly in the Sky)
Natural indigo dye, micronized pure silver, graphite, Hahnemühle paper
8.5 × 11 in (21.59 × 27.94 cm)
2021

0336_Tsukimizutsuki (May, the Month When Rain Makes the Moon Invisible)
Natural indigo dye, micronized pure silver, graphite, Kozo paper
10.5 × 9.5 in (26.67 × 24.13 cm)
2021

0340_Yukigeame (Spring Rain That Melts the Snow)
Natural indigo dye, micronized pure silver, graphite, Kozo paper
10.5 × 9.5 in (26.67 × 24.13 cm)
2021

0349_Niwatazumi (Heavy Rainfall)
Natural indigo dye, micronized pure silver, graphite, Hahnemühle paper
11 × 8.5 in (27.94 × 21.59 cm)
2021

0372_Ame no Inori (Rain Prayer / Praying to the Gods and the Buddha for Rain)
Natural indigo dye, micronized pure silver, graphite, Hahnemühle paper
11 × 8.5 in (27.94 × 21.59 cm)
2021

0377_Kirisame (Fine Rain like Fog and Mist)
Natural indigo dye, micronized pure silver, graphite, Hahnemühle paper
11 × 8.5 in (27.94 × 21.59 cm)
2021

0391_Tade No Ame (Rain Falling on Buckwheat)
Natural indigo dye, micronized pure silver, graphite, Hahnemühle paper
11 × 8.5 in (27.94 × 21.59 cm)
2021

0431_Shūu (Cold Rain in Autumn)
Natural indigo dye, micronized pure silver, graphite, Hahnemühle paper
11 × 8.5 in (27.94 × 21.59 cm)
2021

0455_Taiu tokidoki Furu (Great Rains Sometimes Fall, Season Thirty-six in
Seventy-two Kō Ancient Calendar)
Natural indigo dye, micronized pure silver, graphite, Hahnemühle paper
11 × 8.5 in (27.94 × 21.59 cm)
2021

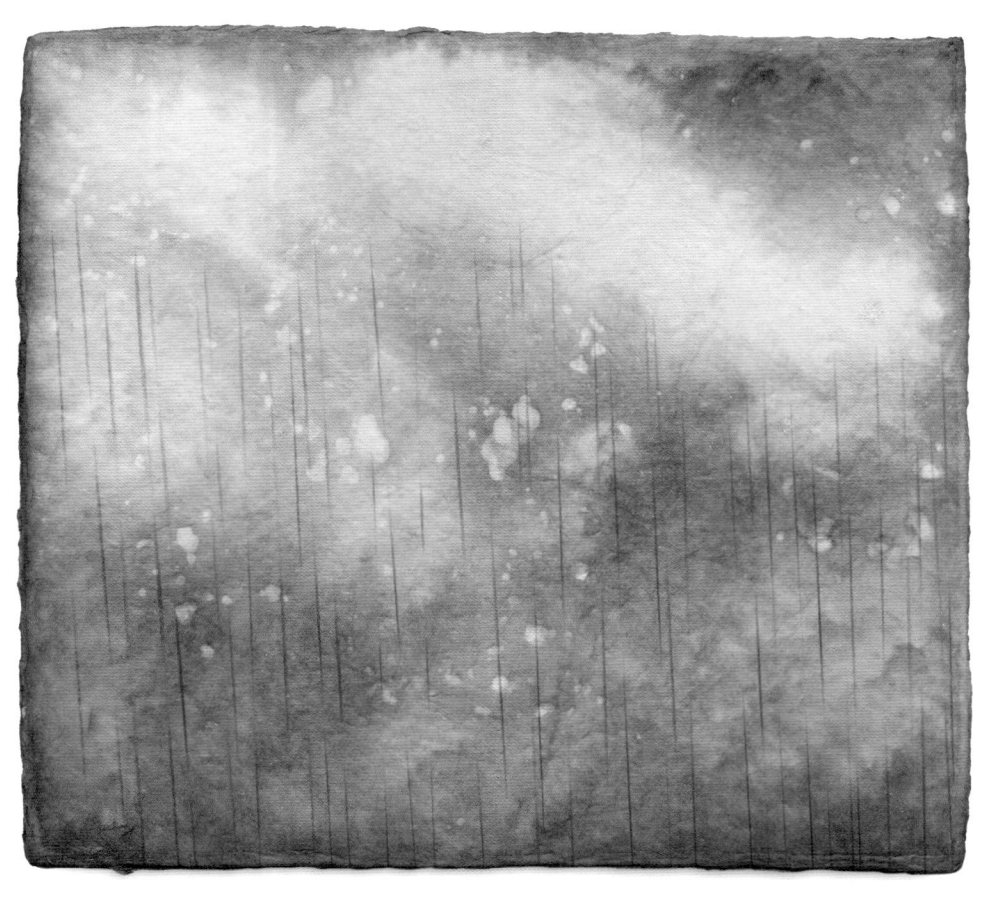

0462_Tamoto no Shigure (Rain on Sleeves / A Kimono Sleeve Wet from Wiping Away Tears)
Natural indigo dye, micronized pure silver, graphite, Kozo paper
9.5 × 10.5 in (24.13 × 26.67 cm)
2021

0480_Rinrin (Continuous Rain for a Long Period of Time)
Natural indigo dye, micronized pure silver, graphite, Hahnemühle paper
11 × 8.5 in (27.94 × 21.59 cm)
2021

0481_Kusa no Ame (Grass Rain / Rain That Falls on New Leaves and Blades of Grass That Have Sprung Up in Spring)
Natural indigo dye, micronized pure silver, graphite, Kozo paper
10.5 × 9.5 in (26.67 × 24.13 cm)
2021

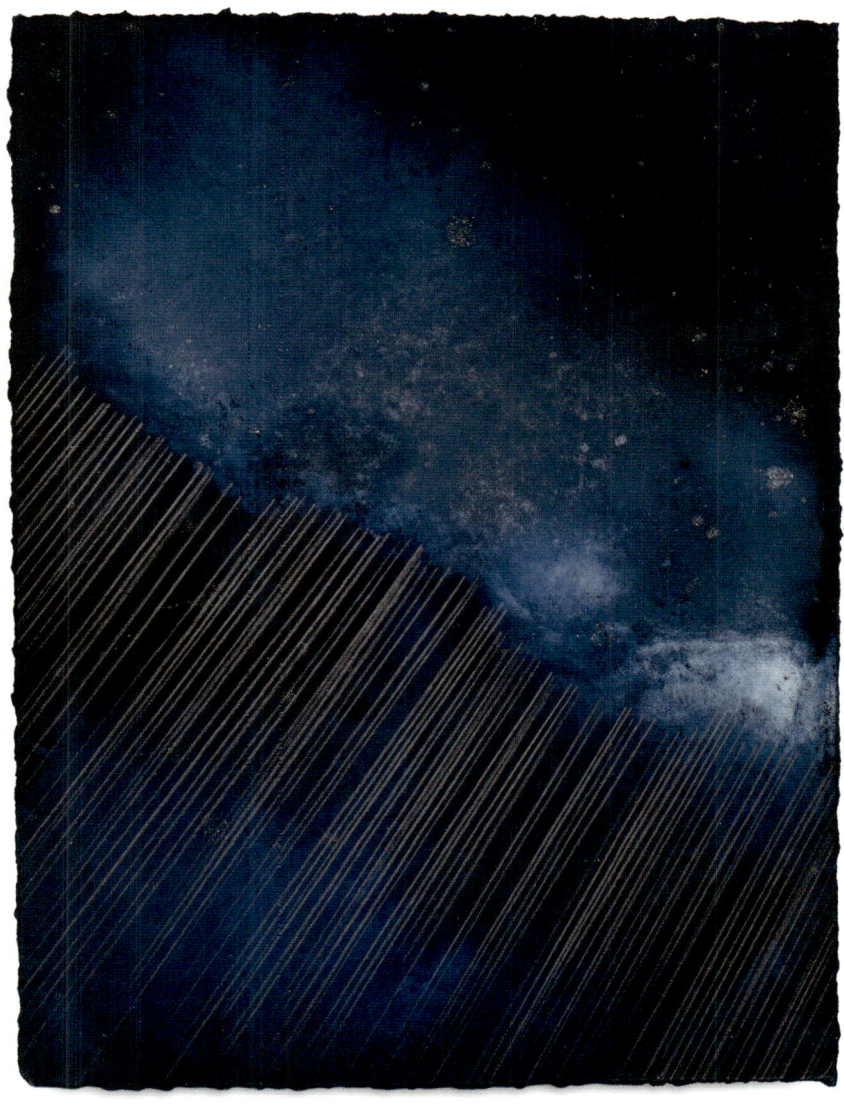

0508_Doshaburi (Pouring Rain All at Once)
Natural indigo dye, micronized pure silver, graphite, Hahnemühle paper
11 × 8.5 in (27.94 × 21.59 cm)
2021

0514_Hanashigure (A Light Rain When the Cherry Blossoms Bloom)
Natural indigo dye, micronized pure silver, graphite, Hahnemühle paper
11 × 8.5 in (27.94 × 21.59 cm)
2021

0515_Kū (Heavy Rain)
Natural indigo dye, micronized pure silver, graphite, Hahnemühle paper
11 × 8.5 in (27.94 × 21.59 cm)
2021

0516_Kuroi Ame (A Black Rain That Fell in Hiroshima Immediately after the Atomic Bomb)
Natural indigo dye, micronized pure silver, graphite, Hahnemühle paper
11 × 8.5 in (27.94 × 21.59 cm)
2021

0525_Uteki (Raindrop)
Natural indigo dye, micronized pure silver, graphite, Hahnemühle paper
11 × 8.5 in (27.94 × 21.59 cm)
2021

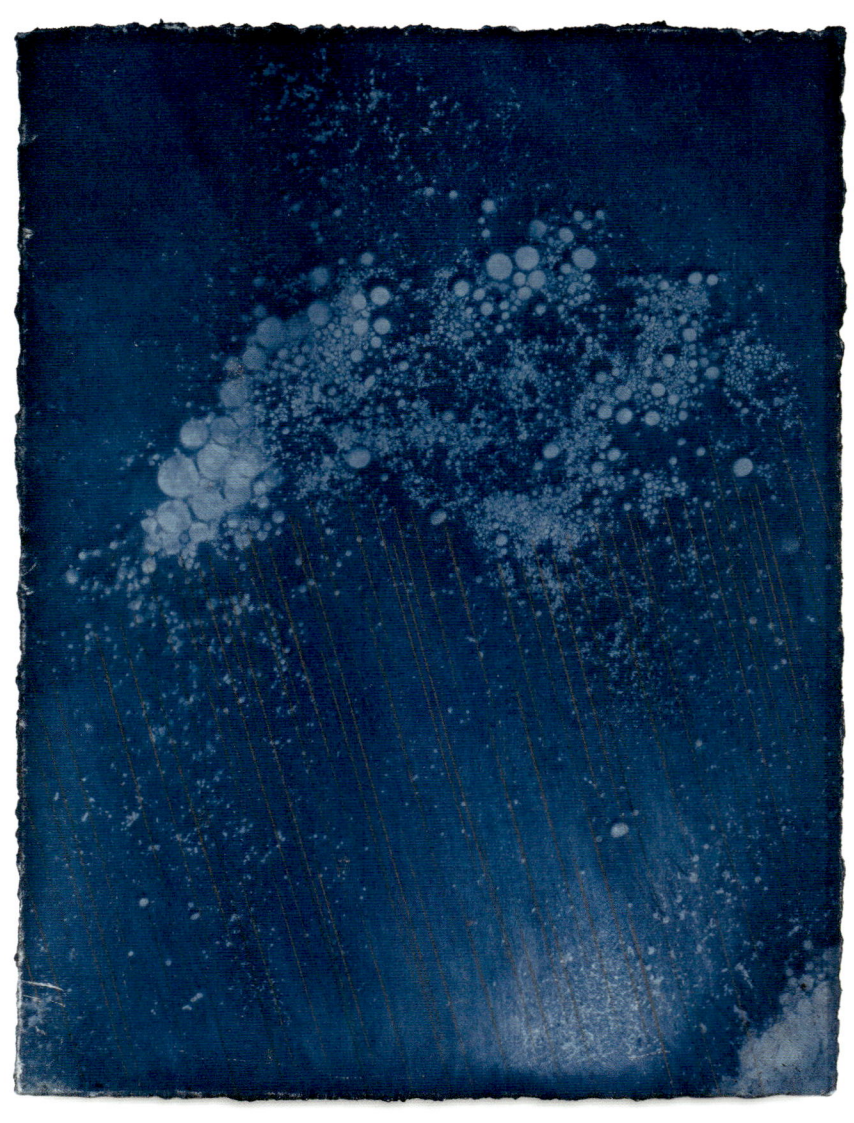

0528_Agarinishiki (A Whimsical Rain—Ishikawa Region)
Natural indigo dye, micronized pure silver, graphite, Hahnemühle paper
11 × 8.5 in (27.94 × 21.59 cm)
2021

0531_Ajisai no Ame (Rain That Falls on and Wets Hydrangea Flowers Beautifully)
Natural indigo dye, micronized pure silver, graphite, Hahnemühle paper
11 × 8.5 in (27.94 × 21.59 cm)
2021

0532_Akebi-kusarashi (Never-Ending Autumn Rain That Rots the Akebi Fruit—
Sado Region)
Natural indigo dye, micronized pure silver, graphite, Hahnemühle paper
11 × 8.5 in (27.94 × 21.59 cm)
2021

0538_Furiguse (To Rain Often as if It Were a Habit / Habitual Rain)
Natural indigo dye, micronized pure silver, graphite, Hahnemühle paper
11 × 8.5 in (27.94 × 21.59 cm)
2021

0544_Aki no Murasame (Autumn Village Rain / Intermittent Rain)
Natural indigo dye, micronized pure silver, graphite, Hahnemühle paper
11 × 8.5 in (27.94 × 21.59 cm)
2021

0570_Rira no Ame (Rain That Falls on Lilac Flowers in a Cold Region near the Sea)
Natural indigo dye, micronized pure silver, graphite, Hahnemühle paper
11 × 8.5 in (27.94 × 21.59 cm)
2021

0595_Ame no Mizu (Rain)
Natural indigo dye, micronized pure silver, graphite, Hahnemühle paper
11 × 8.5 in (27.94 × 21.59 cm)
2021

0638_Shuroshuro (Non-stopping and Gloomy, Weak Rain—Tamana Region, Saitama)
Natural indigo dye, micronized pure silver, graphite, Hahnemühle paper
11 × 8.5 in (27.94 × 21.59 cm)
2021

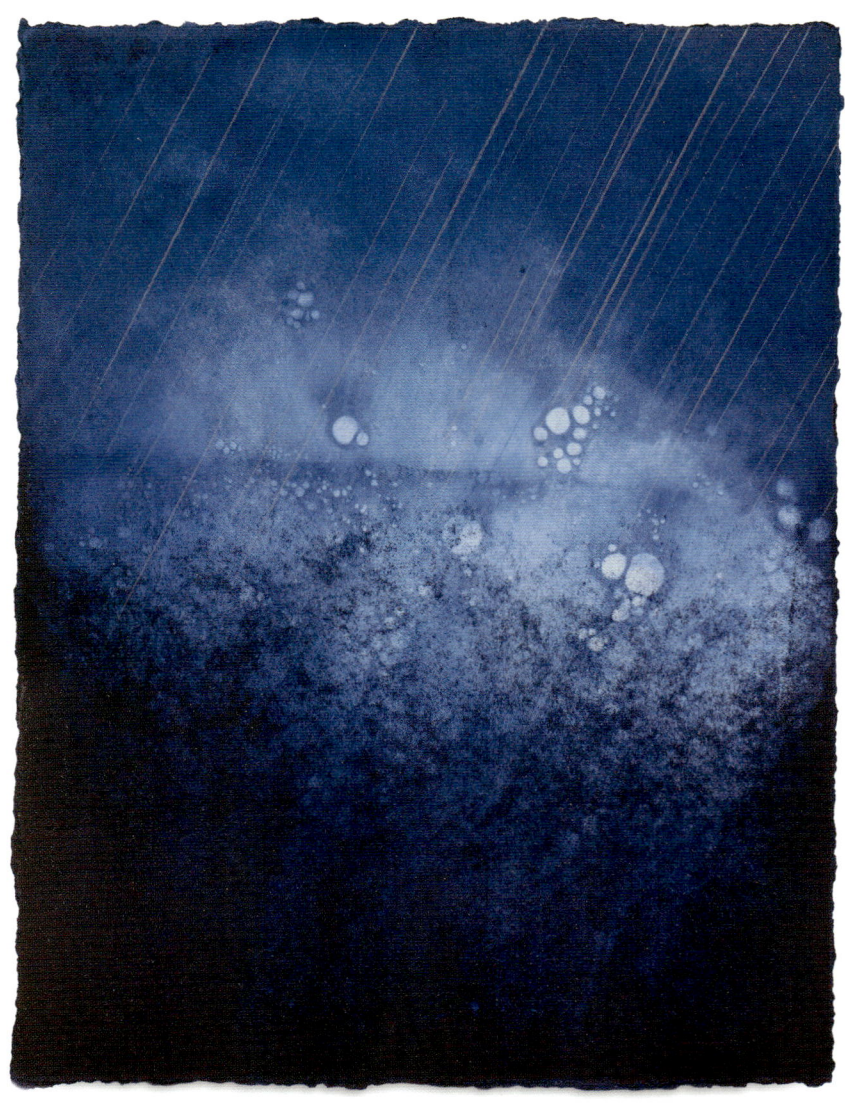

0649_Zen'nyo Ryūō (Female Rain Dragon Deity Who Kūkai Famously Made
Appear during a Rainmaking Contest in 824 at the Kyoto Imperial Palace)
Natural indigo dye, micronized pure silver, graphite, Hahnemühle paper
11 × 8.5 in (27.94 × 21.59 cm)
2021

0653_Kataburi (Raining in One Place but Right Next to It Is Not Raining)
Natural indigo dye, micronized pure silver, graphite, Hahnemühle paper
11 × 8.5 in (27.94 × 21.59 cm)
2021

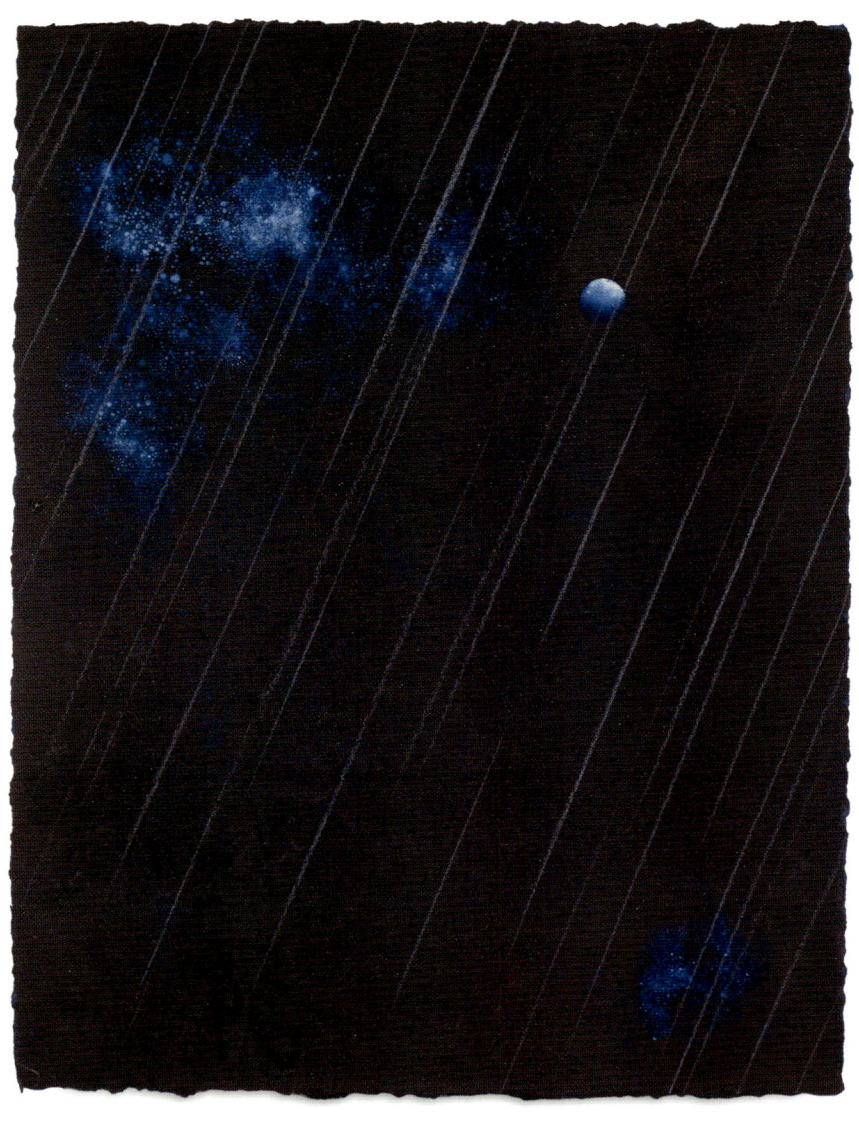

0683_Seigetsu (Moon Shining in a Rain-Cleansed Sky)
Natural indigo dye, micronized pure silver, graphite, Hahnemühle paper
11 × 8.5 in (27.94 × 21.59 cm)
2021

0768_Yarai no Ame (Nonstop and Unrelenting Rain since the Night Before)
Natural indigo dye, micronized pure silver, graphite, Hahnemühle paper
8.5 × 11 in (21.59 × 27.94 cm)
2021

0819_Bōu Senkyū (Rain Chives / To Treat Strangers with Kindness Such as Going Out in the Rain to Pick Chives to Cook Something for Visitors)
Natural indigo dye, micronized pure silver, graphite, Hahnemühle paper
11 × 8.5 in (27.94 × 21.59 cm)
2021

0853_Ameshi o Moyōsu (Rain That Provokes a Feeling Inside to Read Poetry)
Natural indigo dye, micronized pure silver, graphite, Hahnemühle paper
11 × 8.5 in (27.94 × 21.59 cm)
2021

0887_Uki (Praying for Rain)
Natural indigo dye, micronized pure silver, graphite, Hahnemühle paper
11 × 8.5 in (27.94 × 21.59 cm)
2021

0894 Fukuzatsuna Koto o Manabu Yori Mazu, Kaze ya Ame, Yuki ya Tsuki kara Okurarete Kuru Koibumi o Yomu Koto o Oboemashō ("Let Us Learn to Read the Love Letters Sent from the Wind, Rain, Snow, and Moon, Rather than Studying Complicated Things, Such as Sutras" —Ikkyu Sōjun)
Natural indigo dye, micronized pure silver, graphite, Hahnemühle paper
11 × 8.5 in (27.94 × 21.59 cm)
2021

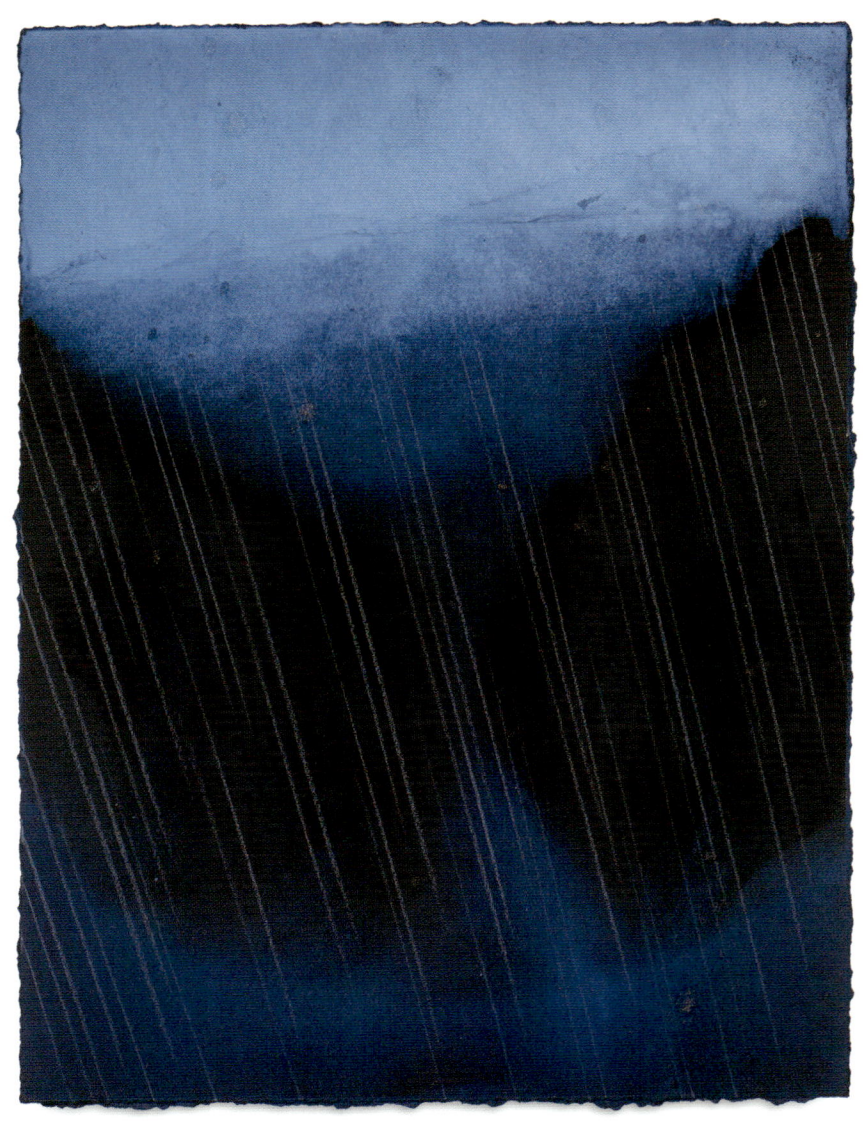

0914_Mishigeru (Mountainous Dialect for Rain—Nishiokitama, Yamagata)
Natural indigo dye, micronized pure silver, graphite, Hahnemühle paper
11 × 8.5 in (27.94 × 21.59 cm)
2021

0975_Sobae (To Rain While the Sun Is Shining / The Marriage of a Fox / Summer)
Natural indigo dye, micronized pure silver, graphite, Hahnemühle paper
8.5 × 11 in (21.59 × 27.94 cm)
2021

1019_Asemokarashi (Brief Summer Rain, like a Sweat Rash That Subsides
When the Temperature Cools—Okayama)
Natural indigo dye, micronized pure silver, graphite, Hahnemühle paper
11 × 8.5 in (27.94 × 21.59 cm)
2021

1061_Hakobiame (A Sudden Rain in the Beginning of Autumn That Is Believed
to Bring the Spirits of Ancestors)
Natural indigo dye, micronized pure silver, graphite, Hahnemühle paper
11 × 8.5 in (27.94 × 21.59 cm)
2021

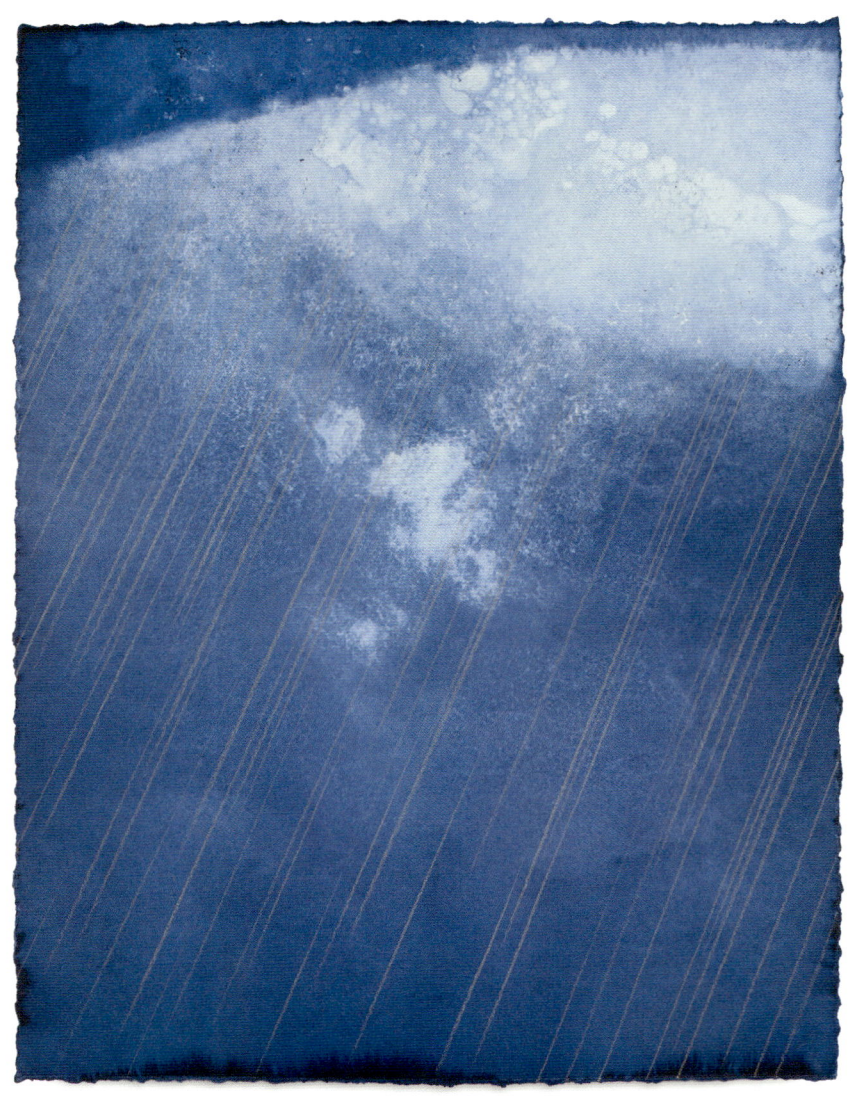

1081_Eishū Gyokuu (Rain as Beautiful as Marble, Falling from the Mountain Where Monks and Deities Live)
Natural indigo dye, micronized pure silver, graphite, Hahnemühle paper
11 × 8.5 in (27.94 × 21.59 cm)
2021

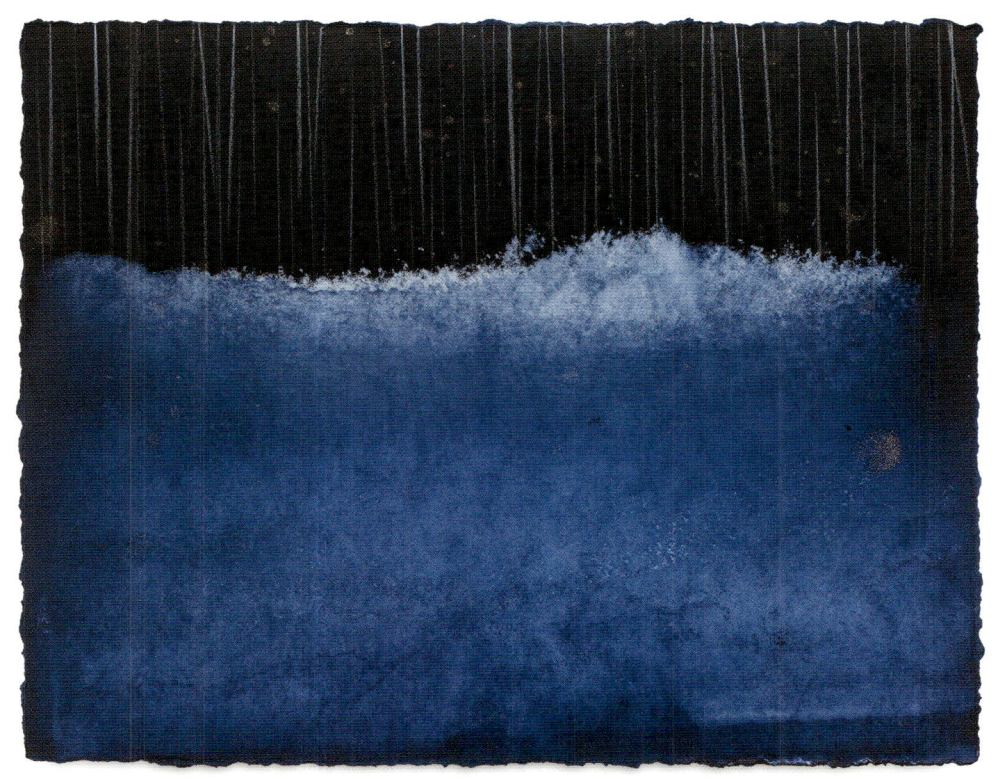

1151_Sau (Rain That Falls on the River Shoal)
Natural indigo dye, micronized pure silver, graphite, Hahnemühle paper
8.5 × 11 in (21.59 × 27.94 cm)
2021

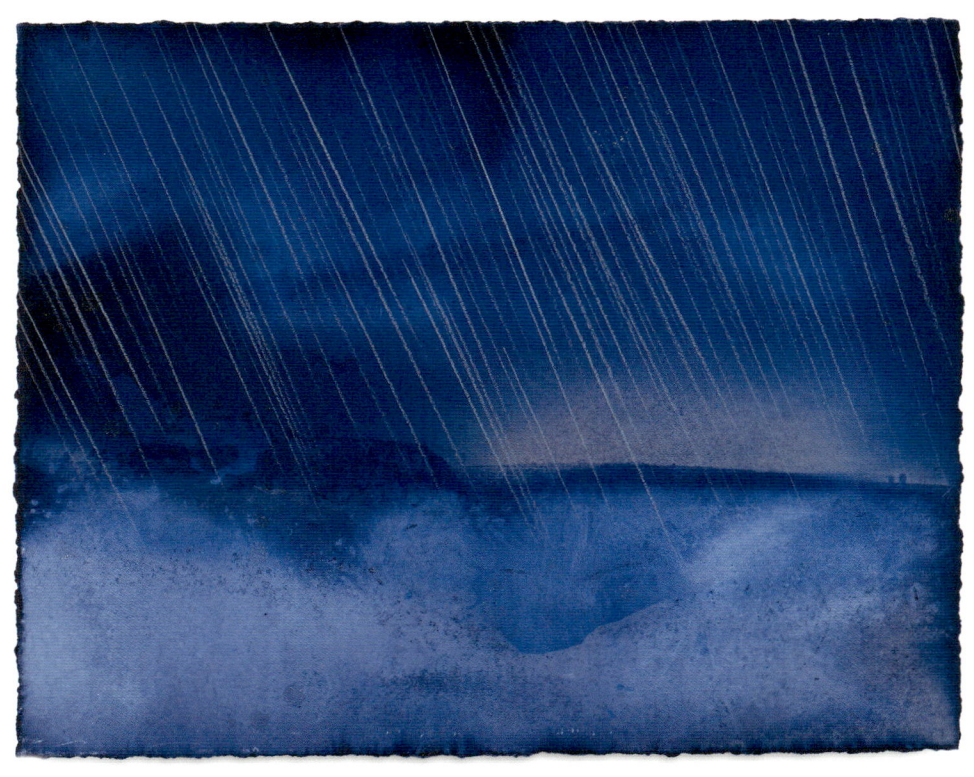

1152_Sanbaine (A Sudden Evening Storm That Occurs So Quickly, One Has
No Time to Make Even Three Bundles of Rice)
Natural indigo dye, micronized pure silver, graphite, Hahnemühle paper
8.5 × 11 in (21.59 × 27.94 cm)
2021

1206_Onfuri (Honorable Downpour Rain on the First Day of the Year That
Blesses the Harvest—January 1)
Natural indigo dye, micronized pure silver, graphite, Hahnemühle paper
8.5 × 11 in (21.59 × 27.94 cm)
2021

1228_Hijiame (A Sudden Rain That Makes One Put Up One's Elbow to Cover One's Head)
Natural indigo dye, micronized pure silver, graphite, Hahnemühle paper
8.5 × 11 in (21.59 × 27.94 cm)
2021

1354_Ushō (To Sing While Being Rained Upon and Wet from the Rain)
Natural indigo dye, micronized pure silver, graphite, Hahnemühle paper
11 × 8.5 in (27.94 × 21.59 cm)
2021

1510_Ame no Yadori (The Lodging of Rain—Metaphor for the Transitory
and the Impermanence of Life)
Natural indigo dye, micronized pure silver, graphite, Hahnemühle paper
11 × 8.5 in (27.94 × 21.59 cm)
2021

1526_Uchū Yurai no Ame (Rain That Comes from Space)
Natural indigo dye, micronized pure silver, graphite, Kozo paper
10.5 × 9.5 in (26.67 × 24.13 cm)
2021

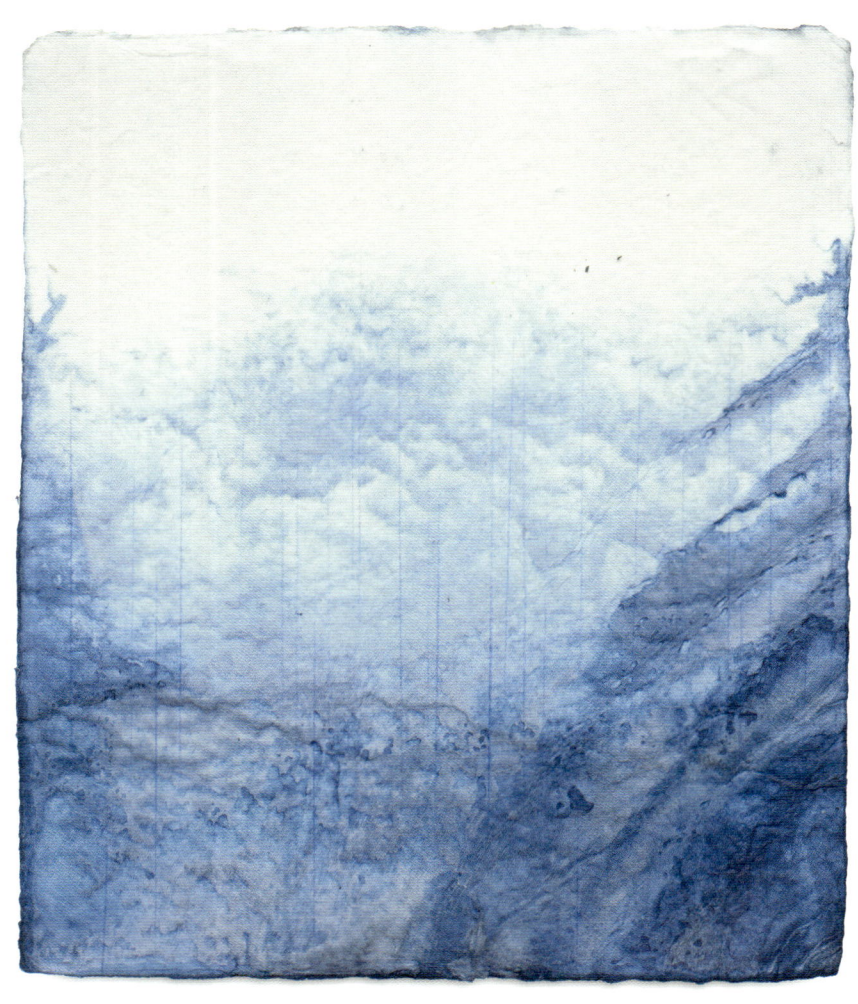

1530_Kokuu (Rain in the Valley)
Natural indigo dye, micronized pure silver, graphite, Kozo paper
10.5 × 9.5 in (26.67 × 24.13 cm)
2021

1608_Tora ga Namida (Tiger Rain on May 28)
Natural indigo dye, micronized pure silver, graphite, Hahnemühle paper
11 × 8.5 in (27.94 × 21.59 cm)
2021

1650_Shigure no Somuru Yama (Mountain-Dyed Rain)
Natural indigo dye, micronized pure silver, graphite, Hahnemühle paper
11 × 8.5 in (27.94 × 21.59 cm)
2021

1716_Tōka no Ame (Rain on Peach Blossoms)
Natural indigo dye, micronized pure silver, graphite, Hahnemühle paper
11 × 8.5 in (27.94 × 21.59 cm)
2021

1720_Takuu (Blessed Rain That Quenches All Things in the Universe)
Natural indigo dye, micronized pure silver, graphite, Hahnemühle paper
8.5 × 11 in (21.59 × 27.94 cm)
2021

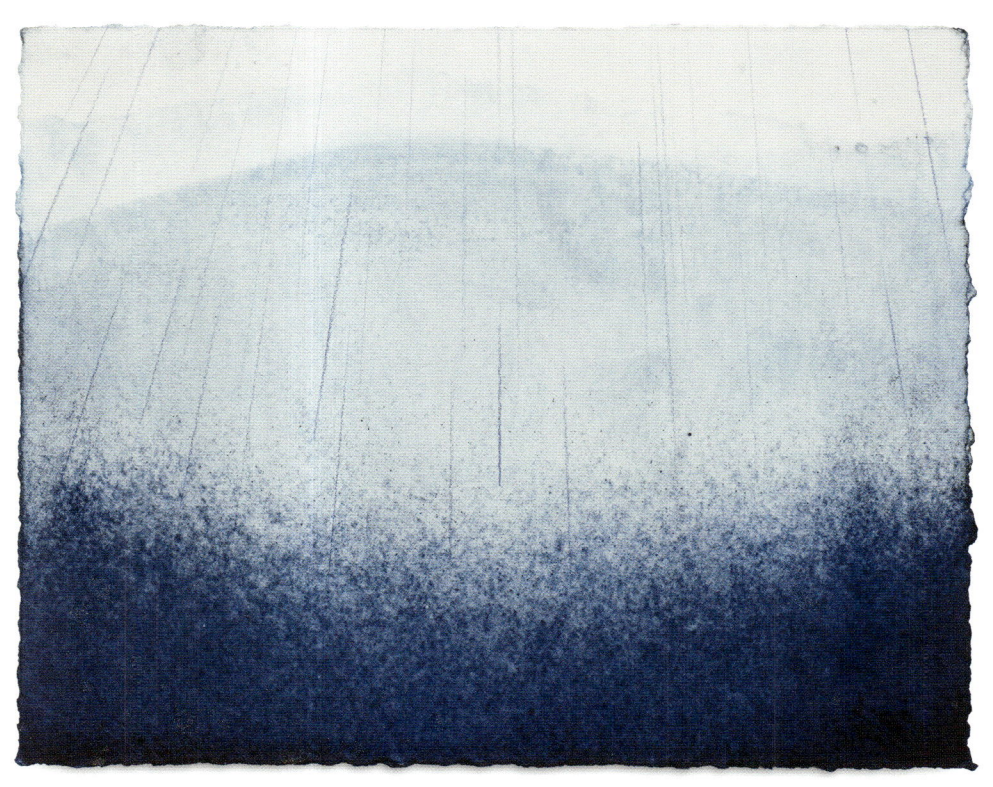

1731_Ki no Meokoshi (Rain That Wakes the Tree Buds)
Natural indigo dye, micronized pure silver, graphite, Hahnemühle paper
8.5 × 11 in (21.59 × 27.94 cm)
2021

1771_Ūamī (Heavy Downpour—Okinawa)
Natural indigo dye, micronized pure silver, graphite, Hahnemühle paper
11 × 8.5 in (27.94 × 21.59 cm)
2021

1826_Ame Sugite Yatō no Aki ("Mysterious Scene of Deep Autumn Colors
on the Pond Even at Night Rain"—Zengo)
Natural indigo dye, micronized pure silver, graphite, Hahnemühle paper
8.5 × 11 in (21.59 × 27.94 cm)
2021

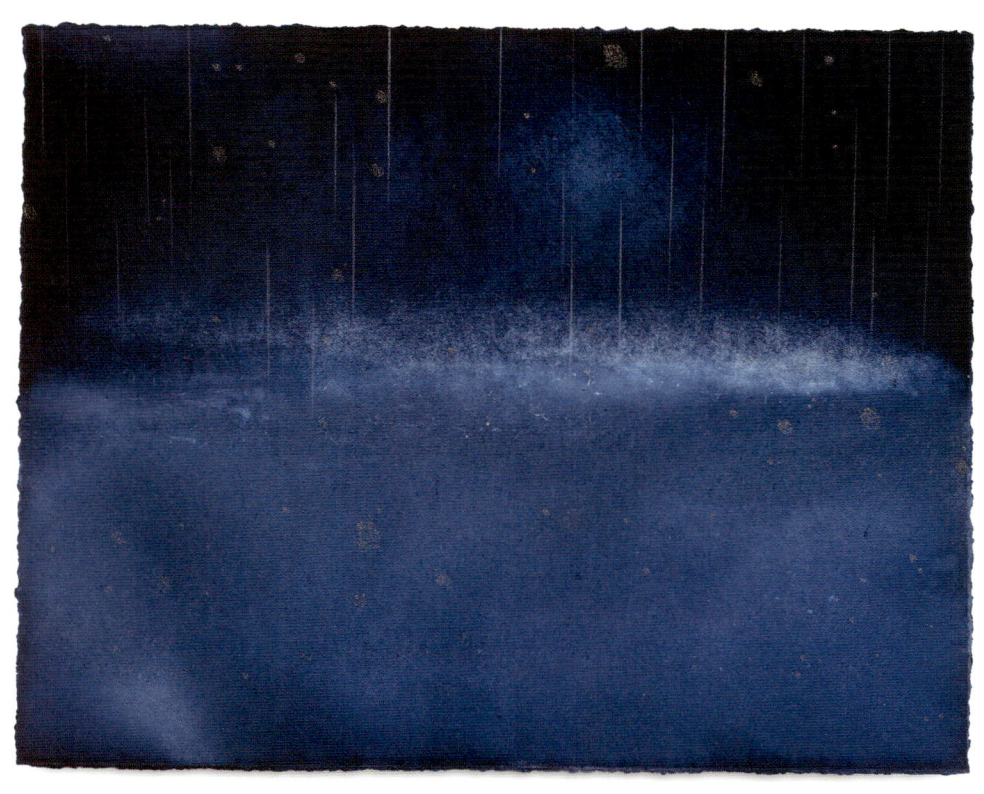

1828_Bai (Spring Dust Rain)
Natural indigo dye, micronized pure silver, graphite, Hahnemühle paper
8.5 × 11 in (21.59 × 27.94 cm)
2021

1836_Rīngaiu (Rain Falling outside the Forest)
Natural indigo dye, micronized pure silver, graphite, Hahnemühle paper
11 × 8.5 in (27.94 × 21.59 cm)
2021

1841_Yōseibaiu (Whimsical Rain That Is Changeable and Could Be Raining Buckets One Moment and Sunny the Next)
Natural indigo dye, micronized pure silver, graphite, Hahnemühle paper
11 × 8.5 in (27.94 × 21.59 cm)
2021

1872_Mienuame (Invisible Rain / Rain So Delicate You Can Hardly See It)
Natural indigo dye, micronized pure silver, graphite, Hahnemühle paper
11 × 8.5 in (27.94 × 21.59 cm)
2021

1987_Shadare (Rain Falling Sideways)
Natural indigo dye, micronized pure silver, graphite, Hahnemühle paper
11 × 8.5 in (27.94 × 21.59 cm)
2021

1988_Uku Wa (Flowers in the Rain / Flowers That Fall in the Rain)
Natural indigo dye, micronized pure silver, graphite, Hahnemühle paper
11 × 8.5 in (27.94 × 21.59 cm)
2021

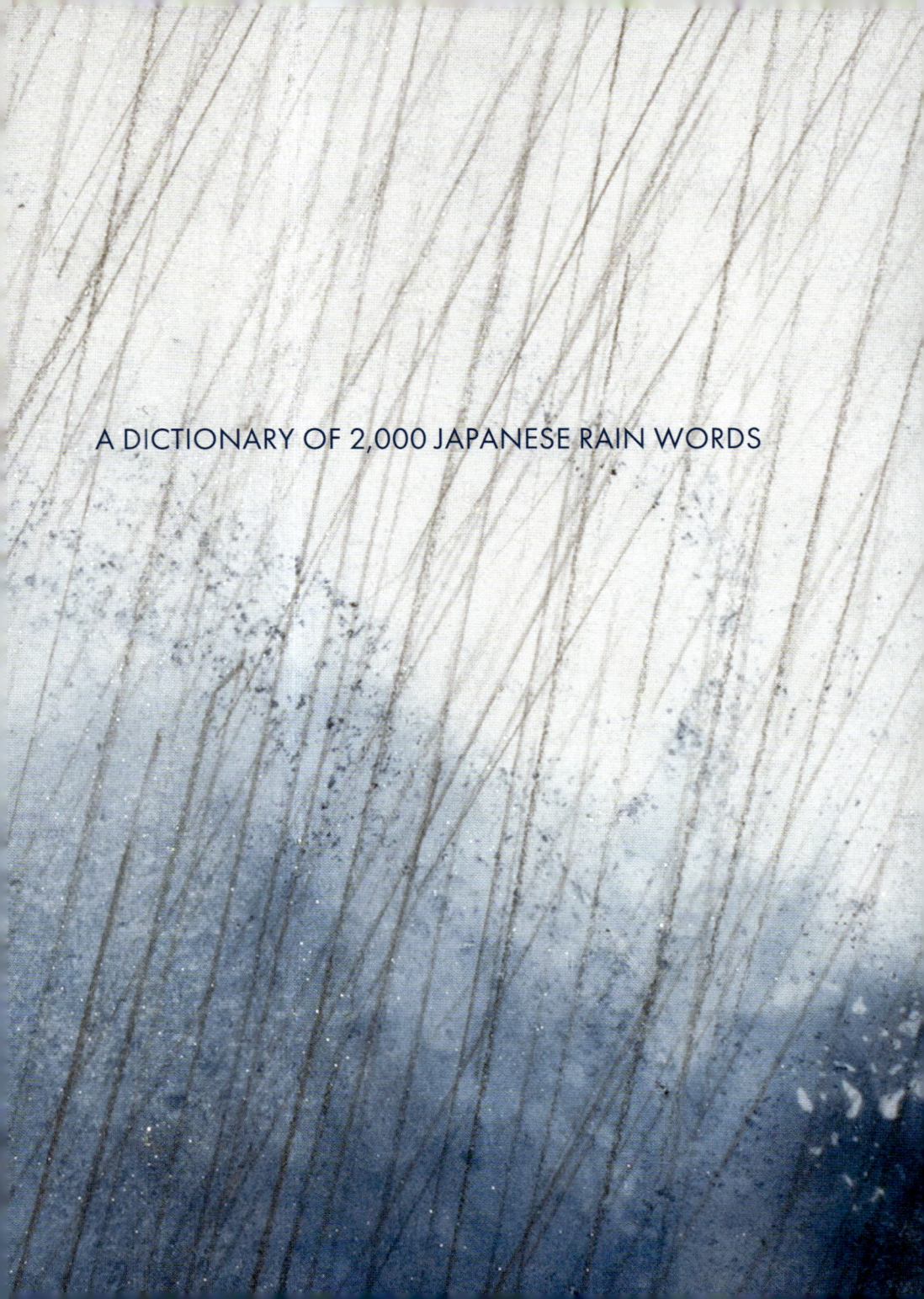

A DICTIONARY OF 2,000 JAPANESE RAIN WORDS

Abaretsuyu (Wild and Heavy Rains Visible at the End of the Rainy Season)	暴れ梅雨	あばれつゆ	0048
Abarezuyu (Rain and Lightning before the End of the Tsuyu Rainy Season)	暴れ梅雨	あばれづゆ	0071
Abe No Seimei (Legendary Wizard Who Was Able to Divine and Call Forth Rain and Whose Mother Was Famously a Fox)	安倍晴明	あべのせいめい	1030
Abura Ame (Oil Rain)	あぶら雨	あぶらあめ	0527
Afurisan (Rainmaking Mountain)	雨降り山	あふりさん	1064
Agarinishiki (A Whimsical Rain—Ishikawa Region)		あがりにしき	0528
Age (Intermittent Rain—Kumamoto Prefecture)		あげ	0529
Agiame (Cold Autumn Rain—Fukushima Prefecture)		あぎあめ	0530
Aiaigasa (Sharing an Umbrella with Someone / Romantic Rain)	相合傘	あいあいがさ	0227
Aiu (Love Rain / Blessed Rain That Falls According to Nature)	愛雨	あいう	0072
Ajisai (Hydrangeas, the Symbol and Quintessential Flower of the Rainy Season)	紫陽花	あじさい	0196
Ajisai no Ame (Rain That Falls on and Wets Hydrangea Flowers Beautifully)	紫陽花の雨	あじさいのあめ	0531
Akebi-kusarashi (Never-Ending Autumn Rain That Rots the Akebi Fruit—Sado Region)	通草腐らし	あけびくさらし	0532
Aki no Ame (Autumn Rain)	秋の雨	あきのあめ	1924

Aki no Ame ga Fureba Neko no Kao ga Sanjaku ni Naru (If It Rains in Autumn, the Cat's Face Will Grow by Three Shaku Because Cats Are Happier on Warm Rainy Days)	秋の雨が降れば猫の顔が三尺になる	あきのあめがふればねこのかおがさんじゃくになる	1603
Aki no Arashi (Autumn Storm)	秋の嵐	あきのあらし	1763
Aki no Jiame (Long-Lasting Heavy Rain during Autumn)	秋の地雨	あきのじあめ	0535
Aki no Murasame (Autumn Village Rain / Intermittent Rain)	秋の村雨	あきのむらさめ	0544
Aki no Nagaame (Long Rain of Autumn)	秋の長雨	あきのながあめ	0331
Aki Shū-u (Light Rain in Autumn)	秋驟雨	あきしゅうう	0278
Akidemizu (Autumn Flood)	秋出水	あきでみず	0540
Akihasan Hongū Akiha Jinja (Shrine in Shizuoka, People Threw Cattle Bones into Sacred Waters and Burned Torches to Irritate and Provoke Kami to Make Rain)	秋葉山本宮秋葉神社	あきはさんほんぐうあきはじんじゃ	1602
Akikosame (Light Rain in Autumn)	秋小雨	あきこさめ	0536
Akikosame (Long Rain in Autumn)	秋微雨	あきこさめ	0001
Akirinu (Long Autumn Rain That Lasts Three Days)	秋霖雨	あきりんう	0546
Akisame (Cold Rain That Falls in Autumn)	秋雨	あきさめ	0002
Akisame Zensen (Autumnal Seasonal Rain Front)	秋雨前線	あきさめぜんせん	0548
Akisazui (Long Rain during Rice-Harvesting Season—Sado Region)	秋さずい	あきさずい	0549
Akishigure (A Shower in Late Autumn)	秋時雨	あきしぐれ	0389

Akishimeiri (A Long Rain in Autumn)	秋湿り	あきしめいり	0274
Akishimeri (A Long Spell of Rain in Autumn)	秋湿り	あきしめり	0266
Akitsuiri (Autumn Rain That Continues like Tsuyu the Rainy Season)	秋微雨	あきついり	0551
Akitsuiri (Autumn Rainy Season)	秋入梅	あきついり	0262
Akitsuiri (Plum Rain / A Long and Continuous Autumn Rain That Is Reminiscent of the Rainy Season)	秋入梅	あきついり	0552
Akitsukiiri (Long Rain in Autumn)	秋黴雨	あきつきいり	0268
Akuten (Stormy, Cold, Bad Weather or Rain)	悪天	あくてん	1766
Ama ga Nukeru (The Skies Open Up, It Rains like Cats and Dogs)	天が抜ける	あまがぬける	0454
Ama Gomoru (Staying Home and Hiding from the Rain)	雨籠	あまごもる	0432
Ama No Gawa Nawashiro-mizu Ni Sekikudase Amakudarimasu Kami Naraba Kami ("Open, O God, the Celestial River, the Milky Way, to Send Water to the Rice Fields"—The Priest Noin's Poem That Was Used as a Rain Prayer)	天の川 苗代水 に せきくだせ あま下ります 神ならば神 （能因法師）	あまのがわ なわしろ みずに せきくだせ あまくだりますかみ ならばかみ（のう いんほうし）	1876
Ama no Mizu (Water from the Heavens / Rain)	天の水	あまのみず	1463
Ama no Yadori (All Things Are Impermanent and Fleeting like the Rain That Falls / Mujō)	雨の宿り	あまのやどり	0042
Amaagari (After the Rain)	雨上り	あまあがり	0304
Amaai (A Pause and Break in the Rain)	雨間	あまあい	0398
Amaake (When the Rain Stops Falling)	雨開	あまあけ	0421

Amaashi (Rain Legs / Streaks of Pouring Rain)	雨足	あまあし	0399
Amaashi (Quick Rain / When Rain Falls with Speed and Passes By Quickly)	雨脚	あまあし	0093
Amaasobi (Rain That Is a Blessing from Above That Comes after a Long Drought / A Joyous Rain)	雨遊び	あまあそび	0542
Amaato (Trace of Rain)	雨跡	あまあと	0553
Amabaki (Flowing Rain)	雨吐き	あまばき	0554
Amabaori (Haori for the Rain)	雨羽織	あまばおり	0267
Amabarashi (The Wind Blows the Clouds after the Rain on a Sunny Day)	雨晴	あまばらし	0424
Amabare (The Skies Clear after Rain)	雨晴	あまばれ	0237
Amabasa (A Place Where Falling Rainwater Goes)	宿吐	あまばさ	0320
Amabashiri (Running Rain / The Edge of Kabuto Helmet)	雨走	あまばしり	0555
Amabata (Rain Field / Black Slate Stone)	雨畑	あまばた	1032
Amabie (It Rains and Becomes Cold)	雨冷え	あまびえ	0311
Amabie (It Rains and Becomes Cold)	雨冷	あまびえ	0543
Amabiki Kannon (Rain-Drawing Goddess of Mercy)	雨引観音	あまびきかんのん	1751
Amabiko (Comes Out in the Rain)	雨彦	あまびこ	0556
Amabiyori (Rainy Weather)	雨日和	あまびより	0212
Amabuta (Cover to Keep from the Rain)	雨蓋	あまぶた	0588

Amadaiko (Rain Drums Used to Shock and Vibrate the Sky and Call Forth Rain)	雨太鼓	あまだいこ	1746
Amadare (Raindrops Falling from the Eaves)	雨垂れ	あまだれ	0161
Amadare Byōshi (Rain Time Signature / Sound of Rain Falling)	雨垂拍子	あまだれびょうし	0321
Amadare Chōshi (Raindrops' Rhythm / In Music an Inexperienced Person Playing Falteringly)	雨垂調子	あまだれちょうし	0557
Amadare Giwa (The Place Where Raindrops Fall)	雨垂際	あまだれぎわ	0558
Amadare Ishi o mo Ugatsu (Raindrops Even a Stone Will Be Pierced Through By)	雨垂れ石をも穿つ	あまだれいしをもうがつ	0088
Amadare no Zensōkyoku (Prelude of Raindrops—Chopin)	雨だれの前奏曲	あまだれのぜんそうきょく	0559
Amadare wa Sanzunokawa (Raindrops Are the River Sanzu / The River of the Afterlife)	雨垂れは三途の川	あまだれはさんずのかわ	0438
Amadareochi (A Place Where Raindrops Fall)	雨垂れ落ち	あまだれおち	0235
Amadareochi (Passing Rain Shower / Falling Raindrops)	雨垂れ落ち	あまだれおち	0265
Amadari (Raindrop or Where the Raindrops Fall)	雨垂り	あまだり	0561
Amadariuke (Place Where Rainwater Flows and Is Collected)	雨垂承	あまだりうけ	0563
Amado (Rain Window)	雨戸	あまど	0568
Amado Mawashi Kanamono (Closure for Rain Window)	雨戸回金物	あまどまわしかなもの	0622
Amadoi (Rain Draining / Sliding Red Beans to Resemble the Sound of Rain)	雨樋	あまどい	0575

Amadoi (To Visit and Check In on Your Neighbors after a Heavy Rain—Kyushu)	雨訪	あまどい	0560
Amadori (Rain Birds)	雨鳥	あまどり	1015
Amadoshi (Rain Year / A Year That Sees More Rain than Usual)	雨年	あまどし	0263
Amaen (Rain Edge outside Door / To Protect from Rain)	雨縁	あまえん	0564
Amagaeru (Rain Frogs That Sing before Evening Showers)	雨蛙	あまがえる	0585
Amagaeru Fukō (Unfaithful Rain Frog—A Boy Who Was Punished and Turned into a Frog That Cries before It Rains for His Misdeeds against His Father)	雨蛙不幸	あまがえるふこう	1752
Amagaeru ga Naku to Ame (If the Rain Frog Croaks, It Will Rain)	雨蛙が泣くと、雨	あまがえるがなくと、あめ	0430
Amagaitō (Rain Cloak)	雨外套	あまがいとう	0586
Amagakari (Where It Is Raining—Miyazaki Region)		あまがかり	0566
Amagake (When It Is Raining, Cover the Kimono)	雨掛	あまがけ	0567
Amagakeru (Soaring Rain / Gods, Souls, Birds Fly in the Sky)	雨翔	あまがける	0332
Amagakure (A Shelter from the Rain)	雨隠	あまがくれ	0565
Amagakure (Hiding from Rain)	雨隠れ	あまがくれ	0210
Amagami (Oil Paper to Protect from Rain)	雨紙	あまがみ	0562
Amagappa (Rain Feather Suit / Oil Coat / Raincoat to Protect from Rain)	雨合羽	あまがっぱ	0607
Amagarami (Rain Rope for Ships)	雨搦	あまがらみ	0589

Amagarami Nawa (Rain Rope for Boats at Sea)	雨搦縄	あまがらみなわ	0640
Amagasa (Rain Hat)	雨笠	あまがさ	0924
Amagasahebi (Rain Umbrella Snake)	雨傘蛇	あまがさへび	0569
Amagasuri (Rain Pattern on a Cloth / A Pattern Showing How It Rained)	雨絣	あまがすり	0439
Amageshiki (Rain Landscape)	雨景色	あまげしき	0579
Amagi (Rain Suit)	雨着	あまぎ	1883
Amaginu (White Silk with Oil Used in Ancient Times to Protect from the Rain)	雨衣	あまぎぬ	0795
Amagiri (A Rain like a Fog / Foggy Rain)	雨霧	あまぎり	0130
Amago (Rain Woman Fish / Salmon during Rainy Season)	雨女魚	あまご	0606
Amagoi (Prayer and Rituals to Invoke Rain)	雨請い	あまごい	0499
Amagoi (Praying for Rain)	雨乞い	あまごい	1216
Amagoi (Rainmaking Prayers)	雨乞	あまごい	0597
Amagoi [Hi o Taite Kemuri de Kumo o Arawasu] (Rainmaking Ritual of Inviting Rain by Burning Fires to Produce Smoke to Look like Clouds)	雨乞い[火を焚いて煙で雲を表す]	あまごい[ひをたいてけむりでくもをあらわす]	0755
Amagoi Jizō (Rain Prayer Jizō—Jizō Sculpture Used in Rainmaking by Tying or Sprinkling with Water)	雨乞い地蔵	あまごいじぞう	1886
Amagoi Komachi (Lady Ono No Komachi, a Poet Who Ended a Drought by Offering a Poem as a Prayer)	雨乞小町	あまごいこまち	1674

Amagoi Komachi (Rain That Fell as a Result of an Imperial Order to Compose a Waka Poem to Pray for Rain)	雨乞い小町	あまごいこまち	0740
Amagoi no Kama (Rainmaking Iron Kettles at Hakone Shrine, Pouring Water from These Kettles on Rice Fields as a Prayer for Bringing Rain)	雨乞いの釜	あまごいのかま	1753
Amagoi no Taki (Pray for Rain Waterfall—Tokushima)	雨乞の滝	あまごいのたき	1643
Amagoi no Tsukai (Messenger Sent to Pray for Rain at Shinsen-En)	雨乞いの使い	あまごいのつかい	0671
Amagoi Odori (A Traditional Dance to Pray for Rain—Hit Kane Bell and Strike Drums and Dance to Beg the Deities and the Buddha for Rain)	雨乞い踊り	あまごいおどり	0675
Amagoi Saijitsu (Rain-Praying Ceremony)	あまごい祭日	あまごいさいじつ	1884
Amagoidake (Praying-for-Rain Mountain, Shiga Prefecture Located in the Suzuka Mountains)	雨乞岳	あまごいだけ	1505
Amagoidaki (Rain Prayer Waterfall)	雨乞滝	あまごいだき	0926
Amagoidori (Rain Bird / It Rains When the Bird of Rain [Red Jade Bird] Cries)	雨乞い鳥	あまごいどり	0613
Amagoihō (Rainmaking Prayer / A Method of Calling the Rain)	雨乞い法	あまごいほう	0375
Amagoiiwa (Rainmaking Rock / Rocks Used in Prayers for Rain)	雨乞岩	あまごいいわ	0492
Amagoimatsuri (Goryu Festival, an Extraordinary Event That Was Held at Shinsen-en Garden to Pray for Rain)	雨乞い祭り	あまごいまつり	1013

Amagoimushi (Rainmaking, Rain Prayer Insect / Tree Frog That Brings Rain)	雨乞い虫	あまごいむし	1014
Amagoiuta (A Song Sung When Begging for Rain / Dancing and Praying for the Rain to Come)	雨乞い歌	あまごいうた	0387
Amagoiuta (Rainmaking Prayer Song to Invoke Rain)	雨乞い唄	あまごいうた	1587
Amagoiwaka (Waka Poem Used as a Prayer for Invoking Rain)	雨乞和歌	あまごいわか	1885
Amagoiyama (Rain Dancing Ritual Mountains / Mountains with Dense Mountain Rain Clouds, Optimal for Rain Dancing, Rainmaking Rituals)	雨乞山	あまごいやま	0581
Amagomori (To Stay Home and Hide from the Rain)	雨籠り	あまごもり	0211
Amagomori Mikasanoyama o Takami Kamo Tsuki no Idekonu Yo ha kutachitsutsu ("High in the Mountains of Mikasa the Moon Is Hidden by Rain"—Man'yōshū [Collection of Ten Thousand Leaves], Volume 6, Song 980)	雨隠り 御笠の山を 高みかも月の出で来ぬ 夜はくたちつつ（万葉集 第六巻 九百八十番歌）	あまごもり みかさのやまを たかみかも つきのいでこぬよは くたちつつ（まんようしゅう だい6かん 980ばんか）	1446
Amagoromo (White Silk Oil Robe for Rain)	雨衣	あまごろも	0811
Amagoshirae (Rain Preparation / Prepared for the Rain)	雨拵	あまごしらえ	0582
Amagu (Rain Tools / All Things Used to Prevent Getting Rained Upon)	雨具	あまぐ	1033
Amagumo (Rain Clouds)	雨雲	あまぐも	0086
Amagumori (Rain Clouds and Overcast Skies)	雨曇り	あまぐもり	0162

Amagurī (Daytime to Evening Rain—Aragusu Island Region, Okinawa)		あまぐりー	0608
Amaguri Higaki (In Years of Rain, Chestnuts Produce Well, In Years of Sunshine, Persimmons Produce Well)	雨栗日柿	あまぐりひがき	0614
Amaguruma (The Sound of Rain / A Prop Used to Express the Sound of Rain in Kabuki / Turning Azuki Beans and Gravel in a Paper and Making the Sound of Rain)	雨車	あまぐるま	0616
Amagusaru (It Rots in the Rain)	雨腐	あまぐさる	0610
Amagyō (Rainmaking / Prayers for Rain)	雨行	あまぎょう	0328
Amahake (The Flow of Rainwater)	雨捌け	あまはけ	0623
Amahare (Rain Ceases and Sky Clears)	雨晴れ	あまはれ	0721
Amaike (Rain Pond / Pond for Collecting Rainwater for Crops)	雨池	あまいけ	0466
Amaiwai (Celebrate Rain after Drought)	雨祝	あまいわい	0628
Amaiwai (To Celebrate by Taking Off a Day from Work When There Is Rain after a Long Period of Sunshine / The Joy of Rain / To Enjoy the Rain)	雨祝	あまいわい	0626
Amajimai (Protection against Rain)	雨仕舞	あまじまい	1684
Amajimeri (To Be Frightened of the Rain / To Be Scared by the Rain)	雨湿り	あまじめり	0472
Amajimi (Rainwater Marks / Stain of Rain)	雨染	あまじみ	0506
Amajitaku (Preparation for the Rain)	雨仕度	あまじたく	1031

Amajitaku (Prepare for Rain)	雨支度	あまじたく	0629
Amakamae (Preparation for Rain)	雨構	あまかまえ	0630
Amakawabari (When You Are in a Procession, Put a Silk or Umbrella Up to Prevent Being Exposed to the Rain)	雨皮張	あまかわばり	0647
Amakawamochi (A Servant Who Manages the Rain Skin during the Procession of the Nobleman)	雨皮持	あまかわもち	0670
Amakawatsuke (Rain Protection for the Oxcart / Rain Skin)	雨皮付	あまかわつけ	0727
Amakaze Shinmatsu (Heavy Wind and Rainstorm—Kagoshima Prefecture Region)	雨風しんまつ	あまかぜしんまつ	0615
Amake (Rainy Scene / Rain Signs / Patterns of Rain)	雨気	あまけ	0631
Amake no Hossan (Rainy Night Stars, Something Rarely Seen)	雨気の星さん	あまけのほっさん	1919
Amakezuku (Notice It Is About to Rain)	雨気付	あまけづく	0213
Amaki no Taiko (Rain Sign Drum—Metaphor for Bad Behavior as Rain Dampens Drums and Muddies Their Sound before It Rains, Which Is a Telltale Sign of Rain)	雨気の太鼓	あまきのたいこ	1877
Amako (Rain Child / Mushrooms)	雨子	あまこ	1892
Amakochi (The Wind and Rain That Blow from the East)	雨東風	あまこち	0617
Amakurotsubame (Rain Bird / Rain Swallow)	雨黒燕	あまくろつばめ	1027
Amakushā (Rain Clouds / Clouds That Bring Rain Any Moment)		あまくしゃー	0778

Amama (While the Rain Has Stopped)	雨間	あまま	0693
Amamado (Rain Window)	雨窓	あままど	0076
Amamayori (A Break from the Rain)	雨間より	あままより	0793
Amamayu (Rain Eyebrow Oxcart)	雨眉	あままゆ	1855
Amamayu no Kuruma (Rain Eyebrow Cart—A Type of Ox Cart with Urushi-E on a White Background for Noble Persons)	雨眉車	あままゆのくるま	0692
Amami (Rain Viewing)	雨見	あまみ	0642
Amamizo (A Small Groove Where Rainwater Falls)	雨溝	あまみぞ	0245
Amamizu (Rainwater)	雨水	あまみず	0376
Amamori (Rain Drips through Leaks / Rain Comes into the House)	雨漏り	あまもり	0632
Amamoya (Rain Mist, a Haze That Lingers after the Rain)	雨靄	あまもや	1754
Amamoyō (Rain Pattern)	雨模様	あまもよう	0580
Amamoyoi (Threat of Rain)	雨催い	あまもよい	1652
Amanjaku (Person Who Seems to Enjoy the Rain / A Type of Demon—Gunma Prefecture)	天邪鬼	あまんじゃく	0633
Amanojaku (A Rain Demon / Rain Devil)	天邪鬼	あまのじゃく	0635
Amaochi (A Place Where Rain-drops Fall from the Roof)	雨落	あまおち	0634
Amaochi Byōshi (Rhythm That Mimics the Raindrops Falling from the Roof, Used When Learning Shamisen)	雨落拍子	あまおちびょうし	0172

Amaochiishi (Rainfall Stone)	雨落石	あまおちいし	0655
Amaōi (Cover Yourself to Protect from the Rain)	雨覆い	あまおおい	0791
Amaosae (Prevent the Rain from Entering)	雨押	あまおさえ	0734
Amaoto (The Voice of Rain / The Sound of Rain)	雨音	あまおと	0270
Amappuri Kazappuki (Rain and Wind / Rain Falls and the Wind Blows)	雨降風吹	あまっぷり かざっぷき	0641
Amaryō (Rain Dragon—A Dragon Who Brings the Rain and Controls the Rain)	雨竜	あまりょう	0351
Amaryū (Rain Dragon)	雨龍	あまりゅう	0192
Amaryū Kantō (Rain Dragon between the Paths—A Traditional Japanese Pattern Based on a Mythical, Spirit Animal)	雨龍間道	あまりゅうかんとう	1360
Amasawari (To Not Go Out for Fear of Getting Wet from Rain)	雨障り	あまさわり	0643
Amashi no Kami (Rain Deities, Prayed to during Drought and to Beg for Rain to Stop during a Flood)	雨師神	あましのかみ	1550
Amashidari (Raindrops)	雨滴り	あましだり	0676
Amashitadari (Raindrops)	雨滴	あましただり	0324
Amashizuku (Raindrops)	雨雫	あましずく	0677
Amashōji (Rain Paper Door / Oil on Shoji to Protect from Rain)	雨障子	あましょうじ	0658
Amasosogi (Drops of Rain / Drizzle)	雨注ぎ	あまそそぎ	0333
Amasōzoku (Prepared for the Rain)	雨装束	あまそうぞく	1217

Amasuji (Rain Whose Trails Are Clearly Visible, Each Droplet Appears to Be Pulling a Thin Thread from the Sky)	雨筋	あますじ	0644
Amatsubame (Rain Birds / Swallows That Are Seen before It Rains)	雨燕	あまつばめ	1016
Amatsubu (Raindrops)	雨粒	あまつぶ	0656
Amatsumizu (Rain or Heavenly Water from Above)	天水	あまつみず	0861
Amatsumizu (Rain)	天津水	あまつみず	1683
Amatsutsumi (Being Trapped by Rain and Unable to Go Outside)	雨障	あまつつみ	0026
Amatsuyu (Rain and Dew)	雨露	あまつゆ	1461
Amatsuzuki (Rain Continues for a Long Time)	雨続	あまつづき	0442
Amauchi (Where the Rain Droplets Fall on the Ground)	雨打	あまうち	1001
Amauchigiwa (Rainfall / A Place Where Rain Falls from the Eaves)	雨打際	あまうちぎわ	0657
Amauke (Rain Catcher)	雨受け	あまうけ	1761
Amauke (Rainwater from the Eaves)	雨承	あまうけ	1739
Amayadori (Taking Shelter from the Rain)	雨宿り	あまやどり	0220
Amayadori Zoku (Rainy Day Shelter Tribe / Taking a Temporary Job with the Intention of Leaving When the Economy Improves)	雨宿り族	あまやどりぞく	1925
Amayami (Waiting for a Pause in the Rain / A Break in the Rain)	雨止み	あまやみ	0712
Amayanashi (A Break in the Rain)		あまやなし	0713

Amayo (Night Rain)	雨夜	あまよ	0015
Amayo no Hoshi (Stars Seen on a Rainy Night, a Very Rare Occurrence)	雨夜の星	あまよのほし	0037
Amayo no Tsuki (Moon on a Rainy Night)	雨夜の月	あまよのつき	0272
Amayōi (Ready for the Rain / Prepared for and Preparation for the Rain)	雨用意	あまようい	0422
Amayoke (Awning to Cover from the Rain)	雨避	あまよけ	1167
Amayorokobi (Rain Celebration, Rejoice in the Rain and Take Time Away from Work to Appreciate the Rain)	雨喜び	あまよろこび	0678
Amayozuki (Moon on a Rainy Night)	雨夜月	あまよづき	1496
Amayuki (Rain and Snow)	雨雪	あまゆき	1002
Amazarashi (Exposed to the Rain / Wet in the Rain)	雨曝	あまざらし	0636
Amazarashi (Leave in the Rain)	雨晒し	あまざらし	0679
Amazareru (It Will Be Damaged by Exposure to Rain / It Will Be Rained Upon)	雨曝れる	あまざれる	0651
Amazaya (Rain Scabbard to Protect Swords and Spears from the Rain)	雨鞘	あまざや	1580
Amazora (Rainy Sky As It Looks When It Is about to Rain)	雨空	あまぞら	0075
Ame (Rain)	雨	あめ	0587
Ame Agaru (It Has Stopped Raining)	雨上がる	あめあがる	0204

Ame Ame Fure Fure Kāsan ga (Rain Rain Fall Fall Mother—Children's Folk Song)		あめあめふれふれ かあさんが	1744
Ame Bon o Katamuku (It Rains So Hard the Tray Tilts Over / It Is Raining Buckets)	雨盆を傾く	あめぼんをかたむく	0547
Ame Chikashi (It Seems like It Will Rain Sometime Soon and the Rain Is Nearing)	雨近し	あめちかし	0697
Ame Furasetamae Dōdotto Dōdotto (Prayer Chanted at the Sky to Call Rain)	雨降らせたまえ ドードットドー ドット	あめふらせたまえ どーどっとどー どっと	1741
Ame Furi Shōgatsu (Celebrate the Rain after a Long Period of Sunshine)	雨降り正月	あめふりしょうがつ	0627
Ame ga Agaru (The Rain Has Stopped)	雨があがる	あめがあがる	0460
Ame ga Furimasu Ame ga Furu (It Will Rain It Will Rain—Children's Folk Song)	雨が降ります雨 がふる	あめがふりますあめ がふる	1740
Ame ga Furō ga Yari ga Furō ga (Whether It Rains or Rains Spears / Strong Determination to Do Whatever It Takes)	雨が降ろうが槍 が降ろうが	あめがふろうがやり がふろうが	0648
Ame ga Futte wa Ikusa ga Deki nu (Cannot Fight the War Because the Battleground Is Rainy)	雨が降っては 戦ができぬ	あめがふっては いくさができぬ	0040
Ame ga Hareru (Rain Passes and Sky Clears)	雨が晴れる	あめがはれる	1926
Ame ga Honburi ni Naru (From Light and Intermittent Rain to Intensified and Continuous Rainfall)	雨が本降りに なる	あめがほんぶりに なる	0685
Ame ga Tōrisugiru (Rain Passes)	雨が通り過ぎる	あめがとおりすぎる	1908
Ame Hare (The Rain Stops Falling, the Sun Comes Out)	雨晴れ	あめはれ	0748

Ame Harete Kasa o Wasuru (When the Rain Stops It Is Easy to Forget an Umbrella / It Is Easy to Forget a Favor during Hardship When Time Has Passed)	雨晴れて笠を忘る	あめはれてかさをわする	0230
Ame Inoru (To Pray for Rain)	雨祈る	あめいのる	1553
Ame Itaru (Starting to Rain)	雨到る	あめいたる	0789
Ame Kemuru (Thin Rain That Looks like Smoke and Blurs the Landscape)	雨煙る	あめけむる	0684
Ame Kizasu (It Will Rain Very Soon)	雨兆す	あめきざす	1605
Ame Majiri (Rain Mixed with Something Else, Snow, Wind)	雨混じり	あめまじり	0410
Ame Majiri (Rain Mixed with Something; Snow, Wind, Sleet)	雨交じり	あめまじり	0790
Ame Nado Furu Mo Okashi ("It Is Charming and Quaint Even When Raining"—*The Pillowbook of Sei Shonagon*, Chapter 1)	雨など降るもをかし（清少納言枕草子 第一段）	あめなどふるもをかし（まくらのそうしだい1だん）	1577
Ame Narazu Shite Hana nao Otsu ("Even If There Is No Rain, Petals Will Fall"—Zengo)	不雨花猶落	あめならずしてはななおおつ	1559
Ame ni Arai Kaze ni Migaku ("Wash in Rain, Polish in Wind"—Zengo)	雨滴聲洗風磨	あめにあらいかぜにみがく	0419
Ame ni Ataru (To Encounter the Rain / To Be Hit by the Rain)	雨にあたる	あめにあたる	1551
Ame ni Au (To Meet the Rain / To Be Caught in the Rain)	雨にあう	あめにあう	0862
Ame ni Kamiarai Kaze ni Kushikezuru (Combing in the Wind and Rain / Hardship as the Storm)	雨に沐い風に櫛る	あめにかみあらいかぜにくしけずる	0652

Ame ni mo Makezu Kaze ni mo Makezu ("I Shall Not Be Defeated by the Rain or Wind" —Kenji Miyazawa)	雨にもマケズ 風にもマケズ（宮沢賢治）	あめにもまけず かぜにもまけず（みやざわけんじ）	0802
Ame ni Nurete Tsuyu Osoroshi Karazu (Those Who Get Wet from the Rain Are Unafraid of the Rainy Season / Those Who Have Encountered Hardships Are Not Afraid)	雨に濡れて露恐ろしからず	あめにぬれてつゆおそろしからず	0772
Ame ni Tsuke Kaze ni Tsuke (Whether It Rains or the Wind Blows, Any Time)	雨につけ風につけ	あめにつけかぜにつけ	1003
Ame no Ashioto (The Sound of Rain Likened to Footsteps)	雨の脚音	あめのあしおと	0857
Ame no Ato wa Jōtenki (After We Have Endured the Rain, We Come Out Stronger When the Sun Shines)	雨の後は上天気	あめのあとはじょうてんき	0687
Ame no Bon (Dancing to Rid Disasters from Rain—Bon Festival)	雨の盆	あめのぼん	0229
Ame no Furu Hi wa Tenki ga Warui (When It Rains, the Weather Is Bad—Metaphor for Saying Something Is Natural)	雨の降る日は天気が悪い	あめのふるひはてんきがわるい	1182
Ame no Hana (Cherry Blossoms in the Rain / Flowers in the Rain)	雨の花	あめのはな	0065
Ame no Hi (A Day of Rain)	雨の日	あめのひ	0698
Ame no Hi-Kun (Mr. Rainy Day / A Man Who Will Come to Pick Up a Woman in His Car on Rainy Days)	雨の日君	あめのひくん	1932
Ame no Hotaru (Fireflies on a Rainy Night)	雨の蛍	あめのほたる	1377

Ame no Inori (Rain Prayer / Praying to the Gods and the Buddha for Rain)	雨の祈り	あめのいのり	0372
Ame no Ito (Rain Thread / Rain That Falls in Knots and Appears like Thread)	雨の糸	あめのいと	0069
Ame no Ka (Fragrance of Rain)	雨の香	あめのか	0669
Ame no Kami (Deity of Rain)	雨の神	あめのかみ	0708
Ame no Kobōzu (A Ghost Child in Kyoto Who Stands Drenched in Front of People's Houses on Rainy Days, People Are Warned Not to Let Him Inside)	雨の小坊主	あめのこぼうず	0889
Ame no Koe (The Voice of Rain / The Sound of Rain)	雨の声	あめのこえ	0006
Ame no Ma (The Continuous Rain Has Stopped for a While)	雨の間	あめのま	0511
Ame no Meigetsu (Mid-Autumn Full Moon Obscured by Rain and Not Clearly Visible)	雨の名月	あめのめいげつ	0335
Ame No Mikajihime No Mikoto (God of Rain)	天御梶日女	あめのみかじひめ	1597
Ame no Miya Kaze no Miya (Rain Shrine Wind Shrine at Ise Jingu)	雨の宮風の宮	あめのみやかぜのみや	0744
Ame no Mizu (Rain)	天の水	あめのみず	
Ame no Oto (The Sound of Rain)	雨の音	あめのおと	0792
Ame no Shiratama (Raindrops That Look like White Balls in Sunlight)	雨の白玉	あめのしらたま	0494
Ame no Tekazu (Trouble Rain / Heavily Pouring and Merciless Rain)	雨の手数	あめのてかず	0761

Ame no ‐suki (Rainy Night Full Moon on the Fifteenth Day of the Eighth Lunar Month, It Rains and the Moon Is Obscured—Autumn)	雨の月	あめのつき	0777
Ame no Tsuyosa (The Strength of Rain / The Intensity of Rain)	雨の強さ	あめのつよさ	0760
Ame no Tsuyu (Rain Dew)	雨の露	あめのつゆ	0493
Ame no Umi (Sea of Rains / Mare Imbrium [A Feature on the Moon])	雨の海	あめのうみ	1882
Ame no Uo (A Famous Rain Fish Saved by Kōbō Daishi / Fish That Are Easy to Catch in the Rain)	鯎	あめのうお	1703
Ame no Jo (Fish of the Rain / Rain Fish Living in Lake Biwa / Trout)	雨の魚	あめのうお	0596
Ame no Yadori (Shelter from the Rain / Rain Inn)	雨の宿	あめのやどり	1778
Ame no Yadori (The Lodging of Rain—Metaphor for the Transitory and the Impermanence of Life)	雨の宿	あめのやどり	1510
Ame no Yamima (The Rain Pauses for a Moment)	雨のやみ間	あめのやみま	0041
Ame no Yō (Like Rain / To Be like the Rain)	雨のよう	あめのよう	1890
Ame Nochi Yuki (Rain Then Snow)	雨後雪	あめのちゆき	0897
Ame o Furasete Korosareta Ryū (The Rain That Killed the Dragon / The Dragon Who Was Killed by the Rain to Save People from the Drought)	雨を降らせて殺された竜	あめをふらせてころされたりゅう	0765
Ame o Kau (To Buy Rain / When It Rains and the Rice Harvest Is Damaged, Buy in Anticipation of a Price Raise)	雨を買う	あめをかう	1881

Ame o Kiku (To Listen Carefully to the Rain / To Hear What the Rain Is Saying)	雨を聴く	あめをきく	0816
Ame o Kīte Kankō Tsuku Mon o Hirakeba Rakuyō Ooshi ("Listen to the Cold Rain, If You Open the Gate There Will Be Many Fallen Leaves"—Zengo)	雨を聴いて寒更尽く 門を開けば 落葉多し	あめをきいて かんこうつく もんをひらけば らくようおおし	1949
Ame o Kīte Kankō Tsuku ("Listening to the Rain and Feeling the Cold"—Zengo)	聴雨寒更盡	あめをきいて、かんこうつく	1518
Ame o Kou (When the Drought Continues, Prayers for Rain)	雨を乞う	あめをこう	0453
Ame o Miru (It Rains / To See the Rain, Observe, View, and Watch the Rain)	雨をみる	あめをみる	0824
Ame o Obitaru Momozakura (Peach Blossom Rain—A Word Used to Describe a Beautiful Woman, as Peach and Cherry Blossoms Look Beautiful When Wet in the Rain)	雨を帯びたる 桃桜	あめをおびたる ももざくら	1888
Ame o Obitaru Tōri (Peach, Plum, and Cherry Blossoms Getting Wet in the Rain)	雨を帯びたる 桃李	あめをおびたる とうり	1627
Ame o Okashi Nira o Kiru (To Go Out in the Rain to Cut Chives in the Garden and Make a Meal for a Friend—Parable of a Deep Friendship)	雨を冒し韮を剪る	あめをおかしにら をきる	1672
Ame o Uru (Selling Rain / Selling Rice in Anticipation of Rain Affecting the Harvest)	雨を売る	あめをうる	0624
Ame Osamarite Sangaku Aoshi (Rainfall Blue Mountain / "The Mountains That Were Obscured by Rain Return to Appearing Blue upon Its Ending"—Zengo)	雨收山岳青	あめおさまりてさん がくあおし	0470

Ame Shajiku no Gotoshi (Heavy Downpour of Rain)	雨車軸の如し	あめしゃじくの ごとし	0291
Ame Shibuku (Rain with Heavy Wind Blowing Sideways)	雨しぶく	あめしぶく	1029
Ame Sōburu (Rain Falls Slowly —Ise Monogatari, Second Stage)	雨そをぶる(伊勢物語 第二段)	あめそをぶる(いさ ものがたり だい 2だん)	0145
Ame Sosogu (Rain Pouring as if a Bucket Was Poured Out)	雨注ぐ	あめそそぐ	0741
Ame Sugite Seitai Uruou ("The Rainfall Made the Moss Shine" —Matsuo Bashō / Zengo)	雨過青苔湿 （芭蕉）	あめすぎて、せい たい、うるおう （ばしょう）	0507
Ame Sugite Yatō no Aki ("Mysterious Scene of Deep Autumn Colors on the Pond Even at Night Rain"—Zengo)	雨過夜塘秋	あめすぎてやとう のあき	1826
Ame to Nari Kumo to Narinikemu, Ima wa Sirazu ("Did You Turn into a Cloud or Rain, I Don't Know" —Murasaki Shikibu, *The Tale of Genji*, Aoi, Chapter 9)	雨となり雲とやな りにけむ、今は知 らず(源氏物語 第九帖 葵)	あめとなりくもとな りにけむ、いまはし らず（げんじものが たり だい9ちょう あおい）	1110
Ame Tsuchikure o Yaburazu (The Rain Falls Quietly and Soaks into the Soil without Destroying It, All the World Is Well and Settled)	雨塊を破らず	あめつちくれをや ぶらず	0818
Ame Tsunoru (Intensifying Rain)	雨募る	あめつのる	0890
Ame Tsuyu o Shinogu (Protect Oneself from Rain and Dew / To Shelter Oneself from the Elements)	雨露を凌ぐ	あめつゆをしのぐ	0533
Ame wa Hana no Hubo (Rain Is the Parent of Flowers)	雨は花の父母	あめははなのふぼ	0116
Ame ya Arare to (Rain and Hail)	雨や霰と	あめやあられと	0261
Ameagari (Immediately after the Rain / Raindrops Are Still Dripping from the Leaves of Trees)	雨上がり	あめあがり	0378

Ameakari (The Sky Appears Faintly Bright Even Though It Is Raining)	雨明り	あめあかり	0016
Amearare (Rain and Hail)	雨霰	あめあられ	0033
Ameashi (Rain That Looks like a Line)	雨脚	あめあし	0362
Ameashi (Rain That Passes By)	雨足	あめあし	0153
Ameato (Just after the Rain Stops)	雨あと	あめあと	0958
Amebare (Rain Stops and Sky Clears)	雨晴	あめばれ	1609
Amedataki (To Be Drenched Outside in the Rain / To Be Struck by Rain—Ama Region, Aichi)		あめだたき	0891
Amefuda (November Rain Suit—Hanafuda)	雨札	あめふだ	1004
Amefurashi (To Send Rain / Sea Hare)	雨虎	あめふらし	0929
Amefuri (Rainfall / To Go to the Police Station, Called Rainfall Because It Will Be a Storm of Trouble)	雨降り	あめふり	1898
Amefuri (Rainfall / Under the Rain)	雨降り	あめふり	0895
Amefuri Hiyadesu (Hyades Star Cluster, the Women Who Make It Rain)	雨降りヒヤデス	あめふりひやです	1414
Amefuri Kazama (Rainy and Windy Day)	雨降風間	あめふりかざま	0404
Amefuri Kozō (A Mythical Spirit of a Samurai Child / A Rain Ghost Child, a Yōkai)	雨降小僧	あめふりこぞう	0731
Amefuri no Taiko (Rainy Drum / It Can't Be Helped—Osaka)	雨降りの太鼓	あめふりのたいこ	1558

Amefuri Nyūdō (Rain Spirit or Ghost That Appears on Rainy Nights as a Haunted, Large Cloud)	雨降入道	あめふりにゅうどう	0817
Amefuri Otsuki San (Rainy Moon—Nursery Rhyme for Children)	雨降りお月さん	あめふりおつきさん	0382
Amefuribana (A Flower That Is Said to Cause Rain if Picked)	雨降花	あめふりばな	0402
Amefuribon (To Celebrate Rain after a Drought)	雨降り盆	あめふりぼん	0945
Amefuriboshi (Rainy Star / The Name of a Star Belonging to One of the 28 Stations)	雨降り星	あめふりぼし	0132
Amefurigusa (Rainfall Grass)	雨降り草	あめふりぐさ	1380
Amefurihana (Rain Flowers / If You Pick Them, It Will Cause Rain)	雨降り花	あめふりはな	1889
Amefuriyama (Rain Mountain / When Clouds and Mist Gather at the Top of the Mountain, It Will Soon Rain)	雨降山	あめふりやま	1212
Amefutte Chijō Uruou (Welcoming Rain That Waters the Crops)	雨降地上湿	あめふってちじょう うるおう	0467
Amefutte Ji Katamaru (Ground That Is Rained On Hardens / Adversity and Difficulties in Life Make One Stronger)	雨降って地固まる	あめふってじかた まる	1757
Amegachi (It Is Rainy)	雨勝ち	あめがち	0900
Amegaeshi (Rain That Returns / Winter Northwest Rainstorm—Akita)	雨返し	あめがえし	0650
Amegeshiki (A Rainy Landscape)	雨景色	あめげしき	0013
Ameguse (Frequent Rain, as if It Has Become a Habit)	雨癖	あめぐせ	0896

Ameichiban (The First Rain That Falls after Risshun without Snow in Northern Japan)	雨一番	あめいちばん	1012
Ameichirei (Rain That Is the Perfect Amount for Farming)	雨一犂	あめいちれい	1034
Ameikka (Over the Rain / After the Rain Has Passed Quickly)	雨一過	あめいっか	0426
Ameitami (Rain Damage / Destructive Rain That Damages Flowers and Other Things)	雨傷み	あめいたみ	0284
Ameiwai (Celebrating the Rain after a Drought by Thanking the Deities)	雨祝	あめいわい	1719
Amejima (Rain Stripe / Traditional Kasuri Pattern)	雨縞	あめじま	1374
Amekaze (Rain and Wind / Driving Rain)	雨風	あめかぜ	0119
Amekaze Shokudō (Rain and Wind Diner / Osaka)	雨風食堂	あめかぜしょくどう	1381
Amekehai (An Atmosphere That Signals That It Is Just about to Rain)	雨気配	あめけはい	1035
Amekingoku (Rain Prison— Emperor Shirakawa Was Prevented from Visiting Hosshōji Temple Three Times Due to Rain, It Also Rained on the Day of His Visit)	雨禁獄	あめきんごく	1214
Amekiri (Rain and Paulownia Card in Hanafuda—Metaphor for Sloppy Play)	雨桐	あめきり	1931
Amekko (Faint and Weak Rain)	雨っこ	あめっこ	1211
Ameko (Light Rain—Aomori)	雨こ	あめこ	1458
Amekoi Jizō (A Bodhisattva Named Jizo Who Loved the Rain)	雨恋地蔵	あめこいじぞう	1620

Amekonkon (Raining—Children's Word)	雨こんこん	あめこんこん	1779
Amemasu (Rain Fish)	雨鱒	あめます	1210
Amemeigetsu (Rainy Night Full Moon)	雨明月	あめめいげつ	1209
Amemeigetsu (Unable to See the Famous Fifteenth Night of the Eighth Lunar Month Moon Due to Rain)	雨名月	あめめいげつ	1335
Amemoyō (Cloudy or Rainy)	雨もよう	あめもよう	0117
Amemoyō (It Looks as though It Will Start Raining Very Soon)	雨模様	あめもよう	1611
Amemoyo (It Seems as though It May Be about to Rain)	雨もよ	あめもよ	1213
Amemoyoi (Impending Rain)	雨もよい	あめもよい	1655
Amenissū (Number of Rainfall Days)	雨日数	あめにっすう	0686
Ameonna (Rain Woman / A Woman Who Brings Rain)	雨女	あめおんな	0490
Ameotoko (Man Who Brings Rain Wherever He Goes)	雨男	あめおとこ	0859
Ameotoko Ameonna (People Who Bring Rain to Important Events)	雨男 雨女	あめおとこ あめおんな	1218
Ameppuri (For It to Rain / Saitama)		あめっぷり	0860
Ameru (To Be Rained Upon)	雨る	あめる	1772
Amesansan (Downpour of Rain)	雨潸々	あめさんさん	0898
Amesettai (Rain Entertainment, Being Hit by Sudden Rain and Viewing the Rain as Omotenashi [Warm and Welcoming Hospitality] from the Heavens)	雨接待	あめせったい	1018

Ameshi o Moyōsu (Rain That Provokes a Feeling Inside to Read Poetry)	雨詩を催す	あめしをもよおす	0853
Ameshikō (Rain and Four Rays of Light—Hanafuda)	雨四光	あめしこう	1776
Ameshinogi (To Withstand the Rain / To Prevent the Rain)	雨凌	あめしのぎ	0066
Ameshitodo (Heavy Pouring Rain)	雨しとど	あめしとど	0700
Ameshizuku (Drops of Rain / Crying)	雨雫	あめしずく	0007
Ameshizuku (Raindrops)	雨雫	あめしずく	0239
Ameshō (I Am Often Rained Upon When I Go Outside)	雨性	あめしょう	1215
Ameshobo (It's Raining)	雨しょぼ	あめしょぼ	1800
Amesobofuru (Tasteful Rain That Falls Steadily, Quietly, Gracefully)	雨そぼふる	あめそぼふる	1325
Amesōjō (Rain Monk Known for His Rain Prayers during the Heian Drought)	雨僧正	あめそうじょう	0852
Ametaifuu (A Typhoon That Has More Rain than Wind)	雨台風	あめたいふう	0976
Ametamore (Rainmaking Prayer Ceremony Conducted in the Kifune Shrine of Kyoto)	雨たもれ	あめたもれ	0858
Ametsubo (Rain Pots Filled with Rainwater / You Will Be Blessed with Rain)	雨壺	あめつぼ	1875
Ametsubu (Raindrop)	雨粒	あめつぶ	0241
Ametsubute (Pebble Rain / A Large Amount of Rain That Falls like Small Stones)	雨礫	あめつぶて	0171
Ametsuyu (Rain and Dew—Metaphor for the Human Condition and Existence)	雨露	あめつゆ	0902

Ametsuzuki (Continuing Rain for a Long Period)	雨続	あめつづき	0392
Ameuba (Rain Nanny / A Ghost That Appears during a Rainy Day, Possibly the Ghost of a Mother Who Lost a Newborn Child)	雨乳母	あめうば	0820
Ameukebana (A Gutter Pointed Upward to Collect Rainwater like a Nose Facing the Sky)	雨承鼻	あめうけばな	0899
Ameuso (Rain Bird / Female Bullfinch)	雨鷽	あめうそ	1037
Ameyasame (Rain and Sharks—Metaphor for Tears and Crying as though It Were Raining)	雨やさめ	あめやさめ	0217
Ameyasame (To Cry One's Heart Out as if Rain Is Pouring Down)	雨や雨	あめやさめ	0043
Ameyashiko (Taking a Break on a Rainy Day)	雨やしこ	あめやしこ	0850
Ameyasumi (To Celebrate by Taking Time Off Work to Rejoice and Celebrate the Rains That Follow a Drought)	雨休み	あめやすみ	0901
Ameyoke (Something Used as a Cover to Protect from the Rain / Shelter from the Rain)	雨除け	あめよけ	0863
Amezutsumi (Wrapped in Rain)	雨づつみ	あめづつみ	1806
An'u (Rain That Falls in the Dark of Night)	暗雨	あんう	0754
Anagura-tsubame (Rain Bird / A Swallow)		あなぐらつばめ	1025
Anekotenki (Sister Weather / Whimsical Rainy Weather—Kazuno Region)	姉こ天気	あねこてんき	1036
Anettai Taurin (Rainforest)	亜熱帯多雨林	あねったいたうりん	0851

Aoarashi (Green Storm / Summer Rainstorm)	青嵐	あおあらし	1819
Aobaame (Rain That Falls in Early Summer on Green Leaves / Blue Leaf Rain)	青葉雨	あおばあめ	0286
Aobashigure (Blue-Green Leaf Rain / When Rainwater Collected by Leaves on a Tree Suddenly Falls like Rain)	青葉時雨	あおばしぐれ	1582
Aoshigure (Blue Rain)	青時雨	あおしぐれ	0260
Aoshigure (Water Drops Falling from Green Leaves)	青時雨	あおしぐれ	0293
Aotsuyu (Blue Rain / Rain That Wets Green Leaves)	青梅雨	あおつゆ	1439
Aoyagi no Yau (Blue Willow Night Rain)	青柳夜雨	あおやぎのやう	0193
Araburi no Ame (Master and Slave Rain / Heavy Rainfall)	主従雨	あらぶりのあめ	1219
Arahae (Wild South Wind at the Beginning of Rain Season)	荒南風	あらはえ	1825
Arakoto (Heavy Wind and Rain—Tottori)		あらこと	1222
Arare (Frozen Raindrops / Hail)	霰	あられ	0205
Araremon (Traditional Hail Pattern)	霰文	あられもん	1820
Arashi no Mae no Shizukesa (The Quiet before a Storm)	嵐の前の静けさ	あらしのまえのしずけさ	1616
Arashi no Makura (Storm Pillow / Sleeping in a Storm)	嵐の枕	あらしのまくら	0188
Arashi no Ue (Top of the Storm)	嵐の上	あらしのうえ	0886
Arashishibori (Traditional Storm Pattern—Aizome)	嵐絞り	あらししぼり	1821

Aratsuyu (Summer Rainy Season)	荒梅雨	あらつゆ	1614
Arazuyu (Wild Rainy Season)	荒梅雨	あらづゆ	0238
Are (Rainstorm)	荒れ	あれ	0214
Ari Ga Su No Deiriguchi O Fusaide Iru To Ame Ga Furu (If You See a Parade of Ants Blocking a Doorway of a Nest, It Will Rain)	蟻が巣の出入口を塞いでいると雨が降る	ありがすのでいりぐちをふさいでいると、あめがふる	1020
Ari no Gyōuretsu o Mitara, Ame (If You See a Parade of Ants, Rain)	蟻の行列を見たら、雨	ありのぎょうれつをみたら、あめ	1038
Asa no Ame (Rain of the Morning)	朝の雨	あさのあめ	1043
Asaame (Morning Rain / Rain That Begins to Fall in the Morning)	朝雨	あさあめ	0745
Asaame ni kasa irazu (Morning Rain Does Not Need Umbrellas)	朝雨に傘いらず	あさあめにかさいらず	1249
Asaame ni kura oke (Put the Saddle on the Horse in Morning Rain / Morning Rain Usually Ends Quickly)	朝雨に鞍置け	あさあめにくらおけ	1822
Asaame Wa Onna No Udema-kuri (Morning Rain Is a Woman's Sleeves Pulled Up / There Is Nothing to Be Afraid Of)	朝雨は女の腕まくり	あさあめはおんなのうでまくり	1701
Asaarashi (Morning Storm—A Sword Made in the Muromachi Period by Ukyo Ryou Shoukou)	朝嵐	あさあらし	1758
Asadachi (Sudden and Intense Rain in the Morning)	朝立	あさだち	1039
Asamadachi (Intense Evening Rain Coming from Asama Mountain)	浅間立	あさまだち	0906
Asamodae (Rain That Falls Only in the Morning—Shimane)	朝もだえ	あさもだえ	1563
Asamodae (Short and Brief Rain in the Morning—Yatsuka Region)		あさもだえ	1044

Asaniji wa Ame, Yūniji wa Hare (Morning Rainbow It Will Rain, Evening Rainbow It Will Be Sunny)	朝虹は雨 夕虹は晴れ	あさにじはあめ ゆうにじははれ	1677
Asashigure (Rain That Falls in the Morning / Winter)	朝時雨	あさしぐれ	0017
Asayake wa Ame, Yūyake wa Hiyori (Dawn Is Rain, Dusk Is Sunshine)	朝焼けは雨、 夕焼けは日和	あさやけはあめ、 ゆうやけはひより	0163
Asemokarashi (Brief Summer Rain, like a Sweat Rash That Subsides When the Temperature Cools—Okayama)	汗疹枯らし	あせもからし	1019
Ashita wa Amefuri Tani'n wa Dorobō to Omoe (Tomorrow It Will Rain, Think of Strangers as Thieves / Proverb)	明日は雨降り他人は泥棒と思え	あしたはあめふり たにんはどろぼうと おもえ	1762
Atatakai Ame (Warm Rain)	暖かい雨	あたたかいあめ	1221
Atozuyu (Late Raining Season)	後梅雨	あとづゆ	0905
Ayashiame (Phantom Monster Rain / Rainfall with Fish, Insects, Sand, Dust Trapped in a Whirlwind)	怪雨	あやしあめ	1220
Babaodoshi (Rain That Frightens Old Women / Sudden Evening Rainstorm That Surprises People with Its Abrupt Arrival—Nagasaki)	婆威し	ばばおどし	0908
Bai (Spring Dust Rain)	霾	ばい	1828
Baika (Rain Season)	梅夏	ばいか	0856
Bairin (Quiet and Fine Rain That Falls around Spring—Another Name for the Rainy Season)	梅霖	ばいりん	0312
Baishiu (Plum Rain, a Seasonal Rain in May)	梅子雨	ばいしう	1073

Baiten (Plum Sky / The Rainy Season)	梅天	ばいてん	0717
Baiu (Rain on Plums / Rainy Season)	黴雨	ばいう	0680
Baiu (The Rainy Season)	梅雨	ばいう	0008
Baiu Kōki (The Latter Half of the Rainy Season, Intense Rainstorms)	梅雨後期	ばいうこうき	0730
Baiu Zenki (The First Half of the Rainy Season, Mild and Prolonged Rain)	梅雨前期	ばいうぜんき	0728
Baiuki (Rainy Season)	梅雨期	ばいうき	0854
Baiuzensen Gōu (Heavy and Strong, Intense Rainstorm at the End of the Rainy Season)	梅雨前線豪雨	ばいうぜんせんごうう	1040
Bakeame (Raining Even Though the Sun Is Shining / Rain in the Sunshine)	化雨	ばけあめ	0716
Baku-u (Rain during the Wheat Season / Rainy Season)	麦雨	ばくう	0345
Ban'u (Rain at Nightfall)	晩雨	ばんう	0752
Banbutsushō (Spring Rain That Provides Life to All That Exists)	万物生	ばんぶつしょう	0910
Bandori-matsuri (Rain That Falls on the Day of a Festival)		ばんどりまつり	0714
Banka no Ame (Rain at the End of Summer, as Autumn Arrives)	晩夏の雨	ばんかのあめ	0715
Banshiu (Calm, Spring Rain That Looks like Tens of Thousands of Thin Needles Suspended from the Heavens)	万糸雨	ばんしう	0855
Barabara (Rain, Rain—Onomatopoeia)		ばらばら	0718
Barabara (Scattering Rain)		ばらばら	0242

Baratsuku (Rain with Larger Drops)		ばらつく	0341
Bashō Yōjō ni Shūnashi ("Melan-choly Rain on Leaves"—Zengo)	芭蕉葉上無愁雨	ばしょうようじょうにしゅううなし	0074
Batsu (Goddess of Drought)	魃	ばつ	1933
Beniame (Red Rain / The Rain That Falls on Flowers and Scatters Their Petals)	紅雨	べにあめ	1286
Benzaiten (The Original Form of a Dragon That Is Prayed to and Petitioned during Rituals for Rain)	弁才天	べんざいてん	1645
Biishiki No Ame (Aesthetics of Rain)	美意識の雨	びいしきのあめ	0457
Bījiunami (Unstable Weather with Rain during Early Spring—Okinawa)		びーじゅんあみ	1287
Bishobisho (Rain That Keeps Falling Incessantly and Continues to Fall)		びしょびしょ	0909
Bishōna Uteki (Rain with Small Raindrops That Is Becoming Weaker)	微小な雨滴	びしょうなうてき	0032
Bisyonure (Soaked with Rain)	びしょ濡れ	びしょぬれ	1829
Biu (To Rain Just a Little Drizzle)	微雨	びう	0160
Bōbi (A Lot of Rain)	滂濞	ぼうび	1288
Bōda (Heavy Rain)	滂沱	ぼうだ	0822
Bōfū Shiu (Rough Wind and Heavy Rains)	暴風駛雨	ぼうふうしう	0315
Bōhai (The Way It Rains Heavily)	滂沛	ぼうはい	0823
Bon No Ame (Rain That Falls around the Time of Bon Festival When Spirits of the Deceased Return to the Home)	盆の雨	ぼんのあめ	0753

Bon'u (Heavy Rain, like an Overturned Container of Water)	盆雨	ぼんう	1041
Bonnagase (Long Rains That Fall near the Time of Spring Equinox—Aichi)	盆ながせ	ぼんながせ	1042
Boro (Light Rain with a Breeze—Chita Region, Aichi)		ぼろ	1173
Bōshūami (Strong Rains during the Raining Season)	芒種雨	ぼーしゅーあみ	1312
Bōsuami (Rainy Season)	ボースアミ	ぼーすあみ	1646
Botabota (The Sound of Raindrops Falling)	ボタボタ	ぼたぼた	0764
Bōtaku (Abundant and Continuous Rain)	滂沢	ぼうたく	1311
Botsubotsu (Sounds of Rain Falling Just a Little Bit)		ぼつぼつ	0123
Bōu (Evening Rain)	暮雨	ぼう	0323
Bōu (Heavy Rain)	暴雨	ぼうう	0871
Bōu (Protection against the Rain)	防雨	ぼうう	0194
Bōu Senkyū (Rain Chives / To Treat Strangers with Kindness Such as Going Out in the Rain to Pick Chives to Cook Something for Visitors)	冒雨剪韭	ぼううせんきゅう	0819
Bunryō no Ame (Sudden, Heavy Rain in June as if Dragons Were Fighting)	分龍の雨	ぶんりょうのあめ	0763
Bunryūu (Rain That Splits a Dragon's Body in Half)	分龍雨	ぶんりゅうう	0955
Buppō to Waraya no Ame wa Dete Kike (Listen to the Rain of Buddhism and Straw Shops / Go Outside to Comprehend Both)	仏法と藁屋の雨は出て聞け	ぶっぽうとわらやのあめはでてきけ	0459

Chābui (Rain That Never Stops—Okinawa)	チャーブイ	ちゃーぶい	1649
Chakuhyō (Rain and Sea Spray Freezing)	着氷(船体着氷)	ちゃくひょう(せんたいちゃくひょう)	0159
Chakuu (Clothing Rain / When It Rains on Your Clothes and Gets You Wet)	著雨	ちゃくう	1382
Chi no Ame (Blood Rain during a War)	血の雨	ちのあめ	1091
Chiame (Mist-like Rain—Kagoshima Region)		ちあめ	0821
Chihayafuru Kami mo Mimasaba Tachisawagi Ama no Togawa no Higuchi Ake Tamae ("If There Are Gods, Please Let Them Open the Celestial Spring"—Ono No Komachi, Poem Written in the Heian Period That Served as a Rain Prayer and Brought Rain for Three Days)	千早振る 神も見まさば 立騒ぎ天の戸川の 樋口あけ給え (小野小町)	ちはやふる かみもみまさば たちさわぎ あまのとがわの ひぐちあけたまえ (おののこまち)	1850
Chijikau (Rain That Knows the Good Time / Perfectly Timed Rain)	知時佳雨	ちじかう	1184
Chikamawari (To Rain Continuously and Repeatedly—Tochigi)	近回り	ちかまわり	1075
Chikeiseikōu (Rainfall Caused by Rising Moist Air on the Slopes of Mountains)	地形性降雨	ちけいせいこうう	1313
Chikuu (Rain on Bamboo Shoots)	竹雨	ちくう	0750
Chirasaame (Rain That Can Barely Be Felt and Is like a Mist That Is Hardly Perceived—Akiyama, Nagano)		ちらさあめ	1132
Chiu (To Perceive the Rain before It Begins)	知雨	ちう	1076

Chō-u (Listen to the Rain)	聴雨	ちょうう	0141
Chōryū (Long Haze Rain)	長霤	ちょうりゅう	1077
Chōunbou (Morning Cloud and Evening Rain)	朝雲暮雨	ちょううんぼう	0134
Chuami (Rainfall)	一雨	ちゅあみ	1680
Chūbyū Miu (Birds Repair Holes in Their Nests before It Rains / To Be Prepared before Disasters Occur)	綢繆未雨	ちゅうびゅうみう	1074
Chūshū Mugetsu (Invisible Full Moon in Mid-Autumn Hidden by Clouds and Rain)	中秋無月	ちゅうしゅうの むげつ	0084
Dai-shō arare (Traditional Pattern Representing Grains of Ice That Fall from the Sky)	大小霰	だいしょうあられ	1637
Daija (Snake Deity Often Venerated in Rain Prayer Rituals)	大蛇	だいじゃ	1592
Daikonzuri (Thin Slices of Radish Sleet—Yatsuka Region, Shimane)	大根摺	だいこんずり	1079
Dairaiu (Very Big and Heavy Thunderstorm)	大雷雨	だいらいう	1080
Daiun Shōgyō (Great Cloud Praying for Rain Sutra—Used by Northern Buddhist Priests to Invoke Rain during a Drought)	大雲請雨經	だいうんしょうう ぎょう	1975
Dan'u (Rain of Bullets / Bullets Falling like Rain)	弾雨	だんう	0751
Dan'u (Warm Rain in Springtime)	暖雨	だんう	1078
Dangan'uchū (Hail of Bullets Falling like a Strong Rainstorm)	弾丸雨注	だんがんうちゅう	1188
Danna Biyori (Husband Weather / Rain That Falls Only at Night and Stops during the Day)	旦那日和	だんなびより	0749

Danzokutekina Ame (Intermittent Rain)	断続的な雨	だんぞくてきなあめ	0477
Dari (Rainfall—Echi Region, Shiga Prefecture)		だり	1083
Datekoki Ame (A Whimsical and Sudden Rain—Niigata Region)	伊達こき雨	だてこきあめ	1185
Deiu (Rain Mixed with Ashes)	泥雨	でいう	0371
Dekkeā'ame (Huge Rain—Akiyama Region, Nagano)		でっけぁあめ	1673
Demizu (The Rivers Flood during Raining Season Due to Heavy Rains)	出水	でみず	1133
Den'u (A Rain Shower with Lightning in Summer)	電雨	でんう	0050
Dōdotto, Dōdotto (Let It Rain, Let It Rain!—Chanted by Farmers during a Rainmaking Prayer Ritual at a Hachiman Shrine during the Taisho Flood)		どーどっと、どーどっと	1663
Dokusho Sanyo (Rain Falling on a Winter Night, the Ideal Condition for Reading)	読書三余	どくしょさんよ	0124
Doroame (Dirt and Muddy Rain)	泥雨	どろあめ	1441
Doshaburi (Pouring Rain All at Once)	土砂降り	どしゃぶり	0508
Doshakeru (Rain That Never Stops—Kaifu, Tokushima)		どしゃける	1682
Doyōame (Raining Often on Saturdays and Sundays)	土曜雨	どようあめ	1289
Doyōame (Soil Rain / Earth Rain / Heavy Rain during Summer)	土用雨	どようあめ	0369
Doyōjike (Rain of the Midsummer)	土用時毛	どようじけ	1082
Doyōme (Rain on Midsummer Sprouts)	土用芽	どようめ	1831

Doyōtsubure (Midsummer Rain That Crushes Soil)	土用潰れ	どようつぶれ	1134
Eishū Gyokuu (Rain as Beautiful as Marble, Falling from the Mountain Where Monks and Deities Live)	瀛州玉雨	えいしゅうぎょくう	1081
Ekikaku (Very Heavy Rain)	エキ霍	えきかく	1084
Ekiu (Rain around November / Liquid Rain)	液雨	えきう	0231
Ekiu (Very Cold, Soaking Rain but Just for a Short Time in Autumn)	液雨	えきう	0279
Ema (Picture Horse—Wishing Tablet Representing Black Horses That Were Offered to the Kami to Bring Rain and White Horses Offered to Stop Rain)	絵馬	えま	1832
Emori (Rain That Leaks through the Umbrella)	柄漏り	えもり	0688
En'u (Misty Rain)	煙雨	えんう	0370
En'u (Rain and Smoke)	煙雨	えんう	1519
En'u (Rain That Falls on Eaves)	簷雨	えんう	0576
En'u (Smoky Rain)	烟雨	えんう	1571
Enryū (Rain Dripping below the Eaves)	簷溜	えんりゅう	0537
Enryū (Raindrops That Drip from the Eaves)	簷霤	えんりゅう	0814
Ezotsuyu (Rainy Season on the Pacific Side of Hokkaido Caused by Cold Winds from the Sea of Okhotsk)	蝦夷梅雨	えぞつゆ	0674
Fēbuyā (Fine, Mist-like Rain— Nakagami Region, Okinawa)		ふぇーぶやー	0625
Fuanntei Shuu (Rain Brought About by Warm Air in the Sky Underneath a Cold Layer of Sky)	不安定驟雨	ふあんていしゅうう	0672

Fubuku (The Wind That Blows the Rain Strongly)	吹雪く	ふぶく	0073
Fuchikunun (Rain That Comes Indoors—Shuri Region)		ふちくぬん	0796
Fuji no Ame (Wisteria Rain / Rain That Falls When Wisteria Flowers Are in Bloom)	藤の雨	ふじのあめ	0140
Fujinoyama Arai (Washing of the Mountain Rain / Strong Rain in Autumn / A Rain That Cleans Mount Fuji)	富士の山洗い	ふじのやまあらい	0125
Fujōnagashi (Rain That Falls after a Festival and Washes All Impurities Away from the Shrine—Tamana, Kumamoto)	不浄流し	ふじょうながし	0959
Fūkanubinn (Rain and Wind Blowing Your Hair / Being Exposed to Harsh Elements While Working)	風鬟雨鬢	ふうかんうびん	1139
Fuki no Ame (Rain That Pours Down on Coltsfoot Leaves)	蕗の雨	ふきのあめ	0539
Fukiburi (Driving Rain and Wind / Storm)	吹き降り	ふきぶり	0280
Fukkake (Whimsical Rain That May Start as Soon as It Stops and Then Starts Again)		ふっかけ	0701
Fukuzatsuna Koto o Manabu Yori Mazu, Kaze ya Ame, Yuki ya Tsuki kara Okurarete Kuru Koibumi o Yomu Koto o Oboemashō ("Let Us Learn to Read the Love Letters Sent from the Wind, Rain, Snow, and Moon, Rather than Studying Complicated Things, Such as Sutras"—Ikkyu Sōjun)	複雑なことを学ぶよりまず、風や雨、雪や月から送られてくる恋文を読むことを覚えましょう（一休宗純）	ふくざつなことをまなぶよりまず、かぜやあめ、ゆきやつきからおくられてくることいぶみをよむことをおぼえましょう（いっきゅうそうじゅん）	0894
Furi (Rainfall)	降り	ふり	0673

Furiakasu (Continuing to Rain until Dawn)	降り明かす	ふりあかす	0541
Furiaru (Wild Rain / Rain That Wildly Pours from the Sky)		ふりある	0797
Furidasu (To Start to Rain)	降り出す	ふりだす	0702
Furiguse (To Rain Often as if It Were a Habit / Habitual Rain)	降り癖	ふりぐせ	0538
Furikomeru (It Rains and Snows So Much, One Cannot Go Out)	降り込める	ふりこめる	1047
Furikomeru (Rain Basket / Rain That Envelopes You and Makes You Stay Indoors)	降り籠める	ふりこめる	1687
Furikuramu (When an Area Darkens Due to Rainfall and Rainclouds)	降りくらむ	ふりくらむ	0868
Furikurasu (Rain That Continues All Day Long until Nightfall)	降り暮らす	ふりくらす	0136
Furimasaru (Intensifying Rain, a Typhoon May Be Approaching)	降りまさる	ふりまさる	0694
Furimi Furazumi (Raining Off and On)	降りみ降らずみ	ふりみふらずみ	0695
Furimono (Falling Things—Used in Renga and Haikai to Refer to Snow, Rain, Frost, Dew)	降り物	ふりもの	0865
Furinokosu (Rain That Does Not Fall in a Specific Location / Rain That Misses a Certain Place and Leaves It Dry)	降り残す	ふりのこす	0866
Furishikiru (It Rains Incessantly)	降りしきる	ふりしきる	0696
Furishiku (Rain That Covers the Ground Evenly like Fallen Leaves Covering a Path)	降り敷く	ふりしく	1059

Furishiramu (When the Sky and Area Brighten despite the Rain / Light and Rain)	降りしらむ	ふりしらむ	0762
Furisobotsu (It Is Raining and You Are Soaking Wet)	降り濡つ	ふりそぼつ	0961
Furisosogu (To Continuously Rain Without Stopping / To Downpour)	降り注ぐ	ふりそそぐ	0931
Furisusabu (Sideways Drizzle Rain)	降り荒ぶ	ふりすさぶ	0736
Furitsubo (When It Finally Rains after a Long Drought —Kesen, Iwate)	降壺	ふりつぼ	0864
Furitsunoru (To Rain Harder)	降り募る	ふりつのる	0946
Furitsuzuku (To Continue to Rain)	降り続く	ふりつづく	0087
Furiyamu (The Rain Stops Falling)	降り止む	ふりやむ	1136
Furu (Falls / Rainfall)	降る	ふる	0825
Furu oto ya Mimi mo Sū Naru Ume no Ame ("Listening to the Plum Rain Makes My Ears Sour" —Matsuo Bashō)	降る音や 耳も 酸うなる 梅の雨 （芭蕉）	ふるおとや みみも すうなる うめのあめ （ばしょう）	1514
Furushajiku (Heavy Rain with Droplets as Thick as Axles)	降車軸	ふるしゃじく	1135
Fūsan Uga (The Pain of Working Outdoors or Traveling in the Rain / To Eat and Sleep While Getting Hit by Rain)	風餐雨臥	ふうさんうが	0826
Fūsetsu (Wind with Snow and Rain)	風雪	ふうせつ	0137
Fushigina Ame (A Mysterious Type of Rain)	不思議な雨	ふしぎなあめ	0090
Futokiame (Fat Rain with Large Droplets)	太き雨	ふときあめ	0867
Futtemo Tettemo (Rain or Shine)	降っても照っても	ふってもてっても	0827

Fūu (Rain That Arrives at a Specific Time / Certain Rain)	恒雨	ふうう	0735
Fūu (Rainy with Wind)	風雨	ふうう	0035
Fūu doushū (Strong Winds and Rain / To Experience Hardships Together in the Same Boat)	風雨同舟	ふううどうしゅう	0742
Fūu Seisei (Cold Rain Is Drizzling and the Wind Is Wild)	風雨凄々	ふううせいせい	1326
Fūu Toki Ari (The Perfect Amount of Rain Mixed with a Wind Not Too Strong, Just the Right Balance)	風雨時あり	ふううときあり	0847
Fūun Shokuu (Cloud in the Mountains and Rain in the Mountains)	巫雲蜀雨	ふうんしょくう	0368
Fūunoka (Rain and Wind That Causes a Disaster)	風雨の禍	ふううのか	1323
Fūuseisei (When the Winds Are Strong and the Rain Is Heavy and Cold and Feels Unpleasant)	風雨凄凄	ふううせいせい	0846
Fuyu no Ame (Rain in Winter)	冬の雨	ふゆのあめ	0181
Fuyuame (Winter Rain, like Sleet)	冬雨	ふゆあめ	0128
Fuyuhideri (Drought during the Time of Winter)	冬旱	ふゆひでり	1797
Fuyusame (Winter Rain)	冬雨	ふゆさめ	1052
Fuyushigure (Light Winter Drizzle)	冬時雨	ふゆしぐれ	0036
Fuzan no Ame (Fuzan Mountain Rain / A Man and Woman Meet in a Dream)	巫山の雨	ふざんのあめ	1204
Fuzan no Un'u (Rain Cloud on a Mountain)	巫山の雲雨	ふざんのうんう	1685
Gakuu (Rain in the Mountain)	岳雨	がくう	0848
Gan'u (Rain That Falls on the Rock)	巌雨	がんう	0786

Gantan ni Furu Ame (Rain That Falls on New Year's Day)	元旦に降る雨	がんたんにふるあめ	1050
Geibaiu (Rain during Lunar March, between Spring and the Rainy Season Just as the Plum Flowers Start Blossoming)	迎梅雨	げいばいう	0844
Gekiu (Very Heavy and Intense Rain)	激雨	げきう	1053
Gekkou (Moon Rainbow)	月虹	げっこう	0784
Genhichiame (Long Rain—Agatsuma, Gunma)	げんひち雨	げんひちあめ	0849
Gerira Gōu (Unexpectedly Strong and Sudden Rainstorm / Guerrilla Rain)	ゲリラ豪雨	げりらごうう	0427
Geshi no Ame (Rain on the Summer Equinox)	夏至の雨	げしのあめ	0845
Getsugakuu (Rain That Falls on the First Day of the Month)	月額雨	げつがくう	0379
Gettan-u (Moon Dawn Rain / Rain That Falls on the First Day of the Month)	月旦雨	げったんう	0785
Ginchiku (Silver Bamboo Rain / Rays of Sunshine Reflecting and Refracting Droplets)	銀竹	ぎんちく	0880
Ginga Tōsha (Very Heavy Rain or a Large Waterfall / A Galaxy Descending to Earth)	銀河倒瀉	ぎんがとうしゃ	0180
Ginsen (When Raindrops Look like Silver Arrows)	銀箭	ぎんせん	0759
Giu (False Rain)	偽雨	ぎう	1048
Godōu (Rain That Wets the Aokiri Tree)	梧桐雨	ごどうう	1049

Gofū Jyūu (Wind Blowing Every Five Days and Rain Every Ten Days, Perfect Weather for Farming and the World Is Peaceful and Calm)	五風十雨	ごふうじゅうう	1051
Gōgō (The Sound of Very Heavy Rain)		ごーごー	1688
Goki Araiame (Rain after the Festival That Washes the Ritual Objects)	御器洗雨	ごきあらいあめ	1224
Gōsetsu (Heavy Rain and Snow That May Cause a Disaster)	豪雪	ごうせつ	1093
Goshiki no Ame (A Blessed and Miraculous Rain of Five Colors That Fell from the Sky the Day the Buddha Was Born)	五色の雨	ごしきのあめ	1060
Gōsure (Heavy Rain—Shimane Prefecture)		ごーすれ	1223
Gōu (Dangerous Rain / Very Heavy Rain That May Cause Severe Disasters and Significant Damage)	豪雨	ごうう	0385
Gōu (Heavy Rain That Falls Vigorously)		ごうう	0147
Gozuburi (Heavy Rain—Hiraizumi Region, Iwate)		ごずぶり	0869
Gussha-gussha (Heavy, Intense Rain—Hita Region, Ōita)		ぐっしゃぐっしゃ	1140
Gūu (Rain That One Encounters by Coincidence or Chance)	遇雨	ぐうう	1065
Guzuguzu (The Sound of Drizzling Rain and Raindrops Falling)		ぐずぐず	0870
Guzutsuku (Rain That Keeps Falling On and On)	愚図つく	ぐずつく	1138

Gyō Fū Shun'u (The World Is Peaceful and Calm, like the Wind and Rain)	尭風舜雨	ぎょうふうしゅんう	0085
Gyoshi no Ame (Rain That Comes after a Drought)	御史の雨	ぎょしのあめ	1186
Gyōu (Rain That Falls at Dawn)	暁雨	ぎょうう	0779
Gyūnbō (Stone Shaped like a Cow's Head from Torii River Used Successfully with Chanting of the Heart Sutra to Pray for Rain)	牛んぼう	ぎゅうんぼう	0170
Gyūsekiu (Summer Rain That Pours on One Side of the Sky, While Being Sunny in Another Part of the Sky)	牛脊雨	ぎゅうせきう	1137
Hachidai Ryūō (Eight Dragon Kings of the Lotus Sutra Who Preside over Rain)	八大竜王	はちだいりゅうおう	1207
Hae (The Raining Season— Shima Region, Mie)		はえ	0780
Hageshī Ame (Strong and Intense Rain)	激しい雨	はげしいあめ	0872
Hai (Big Rain)	沛	はい	1141
Hai (Big Rain / Swampy Wet Rain)	霈	はい	0743
Hai (Falling Rain or Snow)	沛	はい	0199
Hairin (Heavy Rain That Lasts for Three Consecutive Days)	霈霖	はいりん	0873
Haitaku (Heavy Rain)	霈沢	はいたく	1225
Haiu (Rain That Pours Down Heavily)	沛雨	はいう	0781
Haizen (Raining Actively / Raining Hard)	沛然	はいぜん	1208
Hakkau (A Rain That Falls Softly and Quietly around April 4)	発火雨	はっかう	0049

Hakkō (White Rainbow Seen in Light Rain or Fog)	白虹	はっこう	1205
Hakobiame (A Sudden Rain in the Beginning of Autumn That Is Believed to Bring the Spirits of Ancestors)	運雨	はこびあめ	1061
Hakudōu (Rain with Big Droplets That Pours under a Sunny Sky)	白撞雨	はくどうう	0593
Hakushūu (Heavy Rain That Falls Intermittently in Autumn)	白驟雨	はくしゅうう	0215
Hakuu (A Sudden Evening Shower in Spring with White Rain)	白雨	はくう	0173
Hana Hiraite Fūu Ooshi (When the Flowers Bloom, There Is Often Rain and Wind)	花発いて風雨多し	はなひらいてふううおおし	0197
Hana no Ame (Rain Falling on Blossoms—On Sakura)	花の雨	はなのあめ	0289
Hanadoki no Ame (Rain during the Time of Flowers / Cherry Blossom Season)	花時の雨	はなどきのあめ	0445
Hanafubuki (Cold Rain during the Time of Cherry Blossoms / Petals Falling like a Snowstorm)	花吹雪	はなふぶき	0505
Hanakutashi (Flower-Rotting Rain—A Temporary Rainy Season from May to Early June)	花腐し	はなくたし	1686
Hanashigure (A Light Rain When the Cherry Blossoms Bloom)	花時雨	はなしぐれ	0514
Hangeame (Midsummer Rain That Falls Eleven Days from Summer Solstice and Will Predict Whether the Year Will Bring a Good Harvest)	半夏雨	はんげあめ	0051
Hangemizu (Half-Summer Water / Midsummer Rain / Flood Caused by Heavy Rain)	半夏水	はんげみず	0600

Hangeshō (Half-Summer End of Rainy Season)	半夏生	はんげしょう	0599
Hangesui (Midsummer Heavy Rain)	半夏水	はんげすい	1705
Hanshakō (Reflection Rainbow)	反射虹	はんしゃこう	1934
Harae (Purification Rite Used in Rain-Praying Rituals)	祓	あられ	1823
Hare Ichiji Kosame (Sunny, Temporary and Brief Light Rain)	晴一時小雨	はれいちじこさめ	0401
Hare Nochi Ame (Clear Then Rain)	晴後雨	はれのちあめ	0386
Harema (A Lull or Pause in the Rain)	晴れ間	はれま	1322
Haru no Ame (Spring Rain, Sometimes Unpleasantly Cold Rain during Springtime)	春の雨	はるのあめ	0601
Haru no Nagaame (When It Rains for an Extended Period of Time Similar to the Rainy Season but around the Springtime Months of March to April)	春の長雨	はるのながあめ	0882
Haru no Shigure (Temporary and Light Rain in Spring)	春の時雨	はるのしぐれ	0884
Haru no Shūu (Afternoon Showers in Spring)	春の驟雨	はるのしゅうう	0440
Harudemizu (Spring Rain That Melts Snow and Causes Rivers to Flood / Prolonged Rain in Spring)	春出水	はるでみず	0602
Haruhayate (Strong Spring Storm)	春疾風	はるはやて	0122
Harumizore (Sleet That Falls in Spring)	春霙	はるみぞれ	0881
Harurin'u (Spring Rain)	春霖雨	はるりんう	0164

Harusame (Fine Rain That Continues to Fall in Spring / Late Spring Rain)	春雨	はるさめ	0275
Harusame Monogatari (*The Tales of Spring Rain*—Ueda Akinari)	春雨物語	はるさめものがたり	1689
Harushigure (A Drizzling Spring Rain That Repeatedly Starts and Stops as Soon as the Sky Begins to Clear)	春時雨	はるしぐれ	1476
Harushūu (Spring Rain / Afternoon Showers in Spring)	春驟雨	はるしゅうう	0405
Haruyūdachi (Evening Rain Shower in Spring)	春夕立	はるゆうだち	0271
Hashiriame (The Sparse and Soft, Faint Rain That Falls before the Main Downpour Begins)	走り雨	はしりあめ	0594
Hashiritzuyu (Running Rainy Season / Rain That Lasts for Several Days before the Rainy Season and Is a Harbinger of the Rainy Season—Tsuyu)	走り梅雨	はしりづゆ	0292
Hassen no Ame ya Atsumaru Kiku no Tsuyu (The Dew of Chrysanthemums Is Gathered in the Eight Rains—Hattori Senpo)	八専の 雨やあつ まる 菊の露 （服部 沾圃）	はっせんの あめや あつまる きくのつゆ （はっとりせんぽ）	1692
Hasu Tsukite Sudeni Ame o Sasaguru no Kasa Nashi (The Lotus Has Already Gone and There Is No Cover from the Rain)	荷盡已無擎雨蓋	はすつきてすでに あめをささぐるの かさなし	0139
Hatsuarashi (The First Storm in the Beginning of Autumn)	初嵐	はつあらし	0297
Hatsushigure (The First Rain in Winter after the Eighth of November)	初時雨	はつしぐれ	0206
Hatsuyūdachi (The First Evening Rainstorm of the Year)	初夕立	はつゆうだち	1226

Haun'u (Heavy Downpour)	破雲雨	はうんう	0590
Haya Ame (Fast Rain)	早雨	はやあめ	0591
Hayaame (An Evening Light Shower)	速雨	はやあめ	0883
Hayasaame (Sudden Heavy Rain)	暴雨	はやさあめ	0334
Hayasame (Fast Rain)	早雨	はやさめ	0276
Hebi ga Ki ni Noboru to Ame (When a Snake Climbs a Tree, Rain)	蛇が木に登ると雨	へびがきにのぼるとあめ	0524
Heishōu (Rain That an Ancient Chinese General Prayed for during a Drought)	平章雨	へいしょうう	0592
Hen'u (Rain That Falls Only in One Area)	片雨	へんう	1072
Hi no Ame (Ice Rain / Hail)	氷の雨	ひのあめ	1241
Hichisame (Rain That Falls in Large Quantities / Big Rain)	大雨	ひちさめ	1067
Hideri Ame (A Sun Shower, Raining While Sunny)	日照雨	ひでりあめ	1324
Hideriame (To Have Rain While the Sun Is Shining / Wedding Day Celebration of Two Foxes Getting Married)	日照り雨	ひでりあめ	0962
Hiderigami (God of Drought)	魃	ひでりがみ	1939
Hideriniame (Rain for Drought / Realization of Something That One Has Waited a Long Time For)	旱に雨	ひでりにあめ	1718
Hifū San-u (Merciless Rain and Wind That Provoke a Feeling of Sorrow)	悲風惨雨	ひふうさんう	1375
Higanbouzu no Ōkesa Nagashi (Rains So Hard That It Washes the Monks' Robes during Equinoctial Weeks)	彼岸坊主の大袈裟流し	ひがんぼうずのおおけさながし	0979

Higanjike (Long Rain Falling before Spring Equinox)	彼岸時化	ひがんじけ	1058
Higanshike (Rain of the Other Side / Rain during the Equinoctial Week)	彼岸時化	ひがんしけ	1891
Hijiame (A Sudden Rain That Makes One Put Up One's Elbow to Cover One's Head)	ひじ雨	ひじあめ	1228
Hijikasaame (Rain That Is So Sudden, You Put Your Elbows over Your Head to Keep Out of the Rain / Elbow Umbrella Rain)	肘笠雨	ひじかさあめ	0447
Hijōni Hageshī Ame (Extremely Heavy Rain)	非常に激しい雨	ひじょうに はげしいあめ	1057
Hikari no Wa (A Circle of Rainbow Light Caused by Fine Rain and Sunlight)	光の輪	ひかりのわ	1054
Hinataame (Rain While the Sun Is Shining)	日向雨	ひなたあめ	1056
Hiru no Ame (Daytime Rain)	昼の雨	ひるのあめ	0977
Hisa-kata No Ame ("Rain from the Heavens"—Man'yōshū [Collection of Ten Thousand Leaves], Volume 16, Song 3837)	ひさかたの雨（万葉集 第十六巻 三千八百三十七番歌）	ひさかたのあめ（まんようしゅう だい16かん 3837 ばんか）	1695
Hisakata no Ame no Furuhi o Tada Hitori Yamaeni Oreba Ibuse Karikeri ("If You Are Alone on the Mountain on a Rainy Day in Hisaka"—Man'yōshū [Collection of Ten Thousand Leaves], Volume 4, Song 769)	ひさかたの 雨の降る日を ただ 独り 山辺にをれば いぶせかりけり（万葉集 第四巻 七百六十九番歌）	ひさかたの あめ のふるひを ただ ひとり やまえに をれば いぶせ かりけり（まんよう しゅう だい4かん 769ばんか）	1481
Hisame ("Long Rainy Spell," from Nihon Shoki, Volume 16: Emperor Buretsu)	大雨 (日本書紀「第十六巻 武烈天皇」)	ひさめ	1142
Hisame (Rainfall)	飛雨	ひさめ	1623

Hisame (Very Cold Ice, Rain, and Hail)	氷雨	ひさめ	0024
Hisoka Ame (A Very Thin, Fine Rain That Falls in the Beginning of Spring)	ひそか雨	ひそかあめ	1066
Hisshuku (Rainy Night Stars)	畢宿	ひっしゅく	0474
Hito Otoshi (A Shower of Rain—Kagoshima)	一落とし	ひとおとし	0316
Hitoame (Rainfall)	一雨	ひとあめ	0456
Hitoame Ichido (One-Degree Rain / A Rain That Drops the Temperature by One Degree Celsius)	一雨一度	ひとあめいちど	0978
Hitokiriame (One Piece [or One Cut] of Rain / Intermittent Rain)	一切雨	ひときりあめ	1227
Hitokumoare (Rain That Becomes Stronger and Intensifies Temporarily—Niigata)	一雲荒れ	ひとくもあれ	1055
Hitome (One Rain)	一雨	ひとめ	0526
Hitootoshi (Just a Little Shower of Rain)	一落し	ひとおとし	1242
Hitoshibori (Water Squeezed Rain / To Rain Heavily for a Period of Time)	一絞り	ひとしぼり	1229
Hitoshigure (Short, Passing Shower)	一時雨	ひとしぐれ	1244
Hiu (Rain Falling Heavily While the Wind Carries It Away)	飛雨	ひう	1327
Hiyoriame (It Rains Calmly, despite Sunshine in the Sky)	日和雨	ひよりあめ	1232
Hō'u (Autumn and Winter Strong Rain)	澎雨	ほうう	1331
Hō'u (Rain That Falls Where There Is Continuous Sunshine, Rain That Moistens and Is a Blessing)	法雨	ほうう	1267

Hō'u (Rain That Is a Blessing / Rain That Moistens and Nurtures All Living Things)	法雨	ほうう	0837
Hōen Dan'u (A Rain of Bullets)	砲煙弾雨	ほうえんだんう	1330
Hokageame (Playful Rain That Starts and Stops and Starts Again—Aomori)	ほかげ雨	ほかげあめ	1108
Hokori Osae no Ame (Rain to Stop Dust / Just a Little Bit of Rain That Falls on Dry Soil after a Drought)	埃おさえの雨	ほこりおさえのあめ	0838
Homachiame (Occasional and Sporadic Rain in a Limited Area)	外持雨	ほまちあめ	1245
Hon'unfukuu (Cloud and Rain, People's Minds and Hearts Change Easily)	翻雲覆雨	ほんうんふくう	1247
Honbon'u (Rain Becoming Stronger and Does Not Seem to Be Stopping Any Time Soon)	翻盆雨	ほんぼんう	0964
Honburi (Rain or Snow That Begins after Noon and Does Not Seem to Stop)	本降り	ほんぶり	1109
Horoshigure (Somewhat Rainy / A Little Bit of Rain)	ほろ時雨	ほろしぐれ	1246
Hoshi ga Chiratsuku to Ame (When the Stars Flicker, Rain)	星がちらつくと雨	ほしがちらつくと あめ	0452
Hoshi ga Detara Tsugi no Hi wa Ame (When the Stars Come Out and Twinkle, We Will See Rain the Next Day)	星が出たら次の 日は雨	ほしがでたらつぎの ひはあめ	0930
Hoshi Hitotsu Mitsuketaru Yo no Ureshisa wa Tsuki nimo Masaru Samidare no Sora Kobori ("May Rain Sky, Finding One Star Is More Joy than Even the Moon" —Kobori Enshū)	星ひとつ 見つけ たる夜のうれしさ は 月にもまさる 五月雨のそら （小堀遠州）	ほしひとつ みつけ たるよのうれしさは つきにもまさる さみ だれのそら（こぼり えんしゅう）	1412

Hoshikuzu no Ame (Stardust Rain)	星屑の雨	ほしくずのあめ	1268
Hosobikiame (Strong Rain as Thick as Hemp Rope—Niigata)	細引雨	ほそびきあめ	1112
Hosokōri (Thin Ice Crystal Rain and Fog Resembling Diamond Dust)	細氷	ほそこおり	0963
Hōsuiki (The Rainy Season)	豊水期	ほうすいき	0839
Hotobashiru (Gushing and Wet, Raining)		ほとばしる	0885
Hotobiru (The Earth Swells with Moisture from Rain)	潤びる	ほとびる	1329
Houchuū (Abundant and Large Amount of Rain)	豊注	ほうちゅう	1230
Hyakkoku Harusame (Spring Rains for Growing Crops)	百穀春雨	ひゃっこくはるさめ	0109
Hyō (Hail, Rain, and a Thunderstorm)	雹	ひょう	1328
Hyōmu (Ice Fog / Fine Ice Crystals Suspended in the Atmosphere)	氷霧	ひょうむ	0184
Hyōshōsetsu (Precipitation Theory Explaining Clouds Becoming Cold Rain)	氷晶説	ひょうしょうせつ	1107
Ichijiame (One-Hour Rain / Temporary and Very Brief Rain)	一時雨	いちじあめ	0840
Ichijin no Ame (Sudden and Pouring Rain)	一陣の雨	いちじんのあめ	1231
Ichijitekini Furu Ame (A Temporary Rain)	一時的に降る雨	いちじてきにふるあめ	1907
Ichimi no Ame (The Buddha's Teachings Should Be Spread to All People, Just as the Rains Moisten the Grass and Trees)	一味の雨	いちみのあめ	0911
Ichiryō Biyori (Indecisive Weather, Raining or Not Raining)	一両日和	いちりょうびより	1399

Ichiu (One Rainfall / A Single Rain / A Trickle of Rain)	一雨	いちう	0999
Ichiu Sanjitsu (Rain for Three Days in a Row)	一雨三日	いちうさんじつ	0980
Ichiu Senzan o Uruosu (Summer Rain That Only Slightly Wets You, Everything Seems Tranquil and Damp / August)	一雨潤千山	いちうせんざんを うるおす	0185
Iju no Hanaarai no Ame (Rain That Washes the Iju Flowers That Bloom in May / Raining Season—Okinawa)	イジュの花洗い の雨	いじゅのはなあらい のあめ	0982
Ike to Kawa Hitotsu ni Narinu Haru no Ame ("The Pond and River Become One in the Spring Rain"—Yosa Buson)	池と川ひとつにな りぬ春の雨 (與謝 蕪村)	いけとかわ ひとつ になりぬ はるのあめ (よさ ぶそん)	1545
Ikeru Kai Aritsuru Saiwaibito no Hikari Ushinau Hi nite Ame wa Sohofuru Narikeri ("On the Day the Beloved's Light Was Lost, the Sky Wept a Quiet and Gentle Rain"—Murasaki Shikibu, The Tale Of Genji, Wakana, Chapter 35)	生けるかひありつ る幸ひ人の、光失 ふ日 にて、雨はほ そ降るなりけり (源氏物語 第三 十五帖 若菜)	いけるかいありつる さいわいびとの、ひ かりうしなうひにて、 あめはほそふるなり けり(げんじものがた り だい35ちょう わかな)	0179
Ikinari no Ame (A Type of Sudden Rain)	いきなりの雨	いきなりのあめ	0101
Ikkoku Biyori (One Stone Rain / Uncertain Weather)	一石日和	いっこくびより	1097
Ikukau (Spring Rain That Encourages Flowers to Bloom)	育花雨	いくかう	0097
Imiami (Thread-like Rain— Nagahama, Okinawa)		いみあみ	0912
In-Utsu (Feeling Sad and Depressed in Rainy Weather)	陰鬱	いんうつ	0935
In'u (Dirty Rain That Damages the Crops)	陰雨	いんう	0248

In'u (Long Rain That Persists)	隠雨	いんう	1158
In'u (Shade Rain That Continues to Fall)	淫雨	いんう	0108
In'un (Shadow Clouds / Rain Clouds)	陰雲	いんうん	1459
Inasagochi (Wind from the Southeast Brings Rain)	いなさ東風	いなさごち	1396
Inazuma (Lightning and Thunder)	稲妻	いなずま	0078
Inin (Long and Very Persistent Rain)	陰霖	いんいん	0983
Inreki Gogatsu ni Furu Kyūna Ōame (Sudden and Heavy Rain in the Fifth Month of the Lunar Calendar)	陰暦五月に降る急な大雨	いんれきごがつにふるきゅうなおおあめ	1250
Inrin (A Long Rain)	陰霖	いんりん	1248
Inrin (Long Rain)	淫霖	いんりん	0468
Inrō (Long Rain That Continues)	淫潦	いんろう	1162
Insei Baiu (Continuous but Not Strong Rain That Brings Cold Days)	陰性梅雨	いんせいばいう	1000
Iō no Ame (Rain That Is Mixed with Pollen and Particles)	硫黄の雨	いおうのあめ	0932
Ippatsuame (Heavy, Single Rain)	一発雨	いっぱつあめ	1235
Ise Kiyome no Ame (Rain That Purifies after the Ritual on the Day after September 7)	伊勢清めの雨	いせきよめのあめ	0107
Isemuradachi (A Sudden Rain That Comes from the Southwest, an Unsettling Feeling from the Direction of Ise)		いせむらだち	1159
Isoshigure (Rain That Falls in the Coast or the Bay)	磯時雨	いそしぐれ	0913

Issa no Ame (Light Rain, Just Enough to Dampen a Straw Raincoat)	一蓑の雨	いっさのあめ	1400
Isshi Hitori Tsuru Kankō no Ame ("One Thread Cold River Rain"—Zengo)	一絲獨釣寒江雨	いっしひとりつる かんこうのあめ	1801
Isshi no Ame (Very Small Amount of Light Rain That Looks like Flour Being Sifted through a Sieve)	一篩の雨	いっしのあめ	1163
Isshō no Ame (A Very Light Rain for a Brief Period)	一霎の雨	いっしょうのあめ	1401
Iteame (Icy Cold Rain in Winter)	凍雨	いてあめ	1172
Ito Hiku Ame (Rain with Trails So Clearly Visible That They Appear to Be Threads Being Pulled from the Sky)	糸引く雨	いとひくあめ	1233
Itomizu (Thread Water / Raindrops That Fall like Threads from the Eaves)	糸水	いとみず	1160
Itosame (Very Fine Rain, as Thin as Thread)	糸雨	いとさめ	1398
Itsuwari no Shigure (False Rain)	偽りの時雨	いつわりのしぐれ	1397
Ittakuu (A Large Amount of Rain That Opens Flower Buds and Helps the Flowers Bloom)	一拆雨	いったくう	1236
Iwashigumo wa Ame (Mackerel Clouds Signal the Rain)	鰯雲は雨	いわしぐもはあめ	0243
Jajaburi (Insanely Raining / Downpour—Kansai Region)		じゃじゃぶり	0100
Jiame (Continuous and Uniform Rain)	地雨	じあめ	0605
Jibotara Ame (Drenching and Dreary Rain of Wakayama City)	じぼたら雨	じぼたらあめ	0380

Jiburi (Strong and Heavy Down-pour of Rain—Yamanashi Region)	じ降り	じぶり	1176
Jikan Kōsuiryō (Time Rain / Counting the Rain by Time)	時間降水量	じかんこうすいりょう	0355
Jikan Uryō (The Amount of Rain by Time)	時間雨量	じかんうりょう	0788
Jin'u (Heavy Rain)	甚雨	じんう	1113
Jin'u (It Suddenly Rains Very Hard)	迅雨	じんう	0874
Jinkō Kōu (Making Rain by Spraying Silver Iodine on Clouds)	人工降雨	じんこうこうう	1143
Jinsui (Water Offered to God or Deities to Make a Vow)	神水	じんすい	1131
Jīra (Thin Rain—Ozu Region, Ehime)		じーら	0603
Jiri (Fog in the Sea / The Fog That Wets the Face like a Mist or Fine Rain)		じり	1253
Jitajita (Heavy and Intermittent Rain—Yamagata)		じたじた	1252
Jiu (Late Fall or Early Winter Rain Shower / Welcome Rain and Beneficial Rain)	時雨	じう	0256
Jiu (Welcome Rain)	慈雨	じう	1444
Jiunoka (The Blessings of Saints and Princes Will Be Given to All People like the Rain Which Falls Just at the Right Time for the Plants and Trees)	時雨之化	じうのか	1129
Jo'u (Thin and Rather Quiet Rain)	徐雨	じょう	0984
Jōgen no Ame (Ojin Begged for This Rain during a Drought)	状元の雨	じょうげんのあめ	1251
Jōu (Good Rain That Brings Life)	上雨	じょうう	0604
Jōu (Rain All the Time, Regularly)	常雨	じょうう	1130

Jūbako Biyori (Layered Box Rain / Long Autumn Rain—Amakusa Region, Kumamoto)	重箱日和	じゅうばこびより	1370
Jukan'u (Raindrops That Fall from the Leaves, Rainwater That Falls from the Branches)	樹冠雨	じゅかんう	1270
Jukanryū (Rain That Drips along-side the Tree Trunk)	樹幹流	じゅかんりゅう	0773
Jun'u (Order Rain / Rain That Falls According to the Season)	順雨	じゅんう	0904
Jun'u (Rain That Is a Blessing and Arrives at the Perfect Time / Rain That Nourishes the Crops)	潤雨	じゅんう	0732
Juu (Rain That Falls at Just the Right Time)	澍雨	じゅう	1174
Jūu (Refreshing Rain Once Every Ten Days)	十雨	じゅうう	0357
Jyakuu (Weak and Gentle Rain)	弱雨	じゃくう	1234
Jyūfūgo'u (Ten Winds, Five Rains / Perfect Weather for Agriculture, Wind Every Ten Days and Rain Every Five Makes for Calm and Peacefulness)	十風五雨	じゅうふうごう	1269
Jyuhōu (Blessed Rain Law / The Teachings of the Buddha Are Rain That Moistens the Earth)	澍法雨	じゅほうう	1618
Kabashira Tateba, Ame (See a Swarm of Mosquitoes, Signal of Rain)	蚊柱立てば雨	かばしらたてばあめ	0358
Kadachi (Rain Shower—Iwate Prefecture)		かだち	0733
Kadafuri (Rain Falling for Months without Stopping—Hachinohe Region, Aomori)	かだ降り	かだふり	1175

Kadohakigurai (A Brief Rain, Just Enough to Settle the Dust in the Front of the Entrance—Tomata, Okayama)	門掃位	かどはきぐらい	0876
Kae (Cold Winter Rain—Mino Region, Shimane)		かえ	0988
Kaeri Tsuyu (A Returning Rain / The Rainy Season Has Ended and Yet It Rains Again)	返り梅雨	かえりつゆ	0356
Kaeru Mekakushi (Frog Covering Eyes Rain / A Soft Spring Rain That Falls around the Start of Spring Farming—Niigata)	蛙目隠	かえるめかくし	0875
Kageni Hini (Reliable Any Time, during Rain or Shine)	陰に日に	かげにひに	1254
Kagetsuu (Rain That Extends into the Next Month and Also the Month After)	過月雨	かげつう	1465
Kaisōu (Rain That Melts Frost to Protect Crops)	解霜雨	かいそうう	0346
Kaiu (Rain That Feels Refreshing after a Period of Sunshine)	快雨	かいう	1464
Kaiu (Strange Mystery Rain / Earth and Sand and Insects Mixed with Rain and Wind)	怪雨	かいう	1445
Kakikurashi Hare Senu Mine no Amagumo ni Ukite Yo o Furu Miomo Nasabaya ("I Want to Be like the Dark Mountain Peak Rain Cloud That Never Clears, I Want to Be the Smoke in the Sky"—Murasaki Shikibu, *The Tale of Genji*, Ukibune, Chapter 51)	かき暮らし晴れせぬ峰の雨雲に浮きて世をふる身をもなさばや(源氏物語 第五十一帖 浮舟)	かきくらしはれせぬみねのあまぐもにうきてよをふる みをもなさばや(げんじものがたり だい51 ちょう うきぶね)	1900

Kakikurashi Shigu Ruru Sora o Nagame Tsutsu Omoikoso Yare Kaminabi no Mori ("Gazing at the Darkening Rainy Sky, I Think about the Forest Where Gods Dwell"—Kino Tsurayuki)	かきくらし 時雨 るる空を ながめ つつ 思いこそや れ 神なびの森 （紀貫之）	かきくらし しぐるる そらを ながめつつ おもいこそやれ かみなびのもり （きのつらゆき）	1706
Kakishigure (Rainy Sky That Suddenly Becomes Dark)	掻時雨	かきしぐれ	1402
Kakitaru (The Sky Darkens from Rainclouds Low in the Sky)	描き垂る	かきたる	1258
Kakutetsuu (Raining on One Side of the Sky While Being Sunny in Another Part of the Sky in May)	隔轍雨	かくてつう	1240
Kamado no Kemuri ga Tanabikeba Ame (When the Smoke Flutters, It Rains)	かまどの煙がたな びけば雨	かまどのけむりが たなびけばあめ	0216
Kaminarigumo (Lightning Clouds That Cause Rain)	かみなり雲	かみなりぐも	1857
Kamisama ni Tatoerareta Ame (Rain Likened to God)	神様にたとえら れた雨	かみさまにたとえら れたあめ	0478
Kamowake-Ikazuchi No Mikoto (Rain and Thunder Deity)	賀茂別雷命	かもわけいかづちの みこと	1879
Kan no Ame (Cold Rain That Falls from the Small Cold to Early Spring)	寒の雨	かんのあめ	0363
Kan'ake no ame (Rain after the Cold, Rain of the Arrival of Spring)	寒明の雨	かんあけのあめ	0987
Kan'u (Praying for Rain)	喚雨	かんう	1256
Kan'u (Sweet Rain / Rain That Gently, Gently Moisturizes the Vegetation)	甘雨	かんう	0393
Kan'u (Winter Cold Rain)	寒雨	かんう	0295
Kan'ukyū (A Pigeon That Brings Rain When It Cries / Rainmaking Bird)	喚雨鳩	かんうきゅう	1203

Kanbatsu (Drought)	旱魃	かんばつ	1940
Kandachi (Sudden Evening Shower with Thunder)	神立	かんだち	0249
Kanku no Ame (Cold Rain in Winter)	寒九の雨	かんくのあめ	0352
Kanku no Ame (Rain That Falls on the Ninth Day after the Beginning of the Cold Weather / Winter)	寒九の雨	かんくのあめ	0182
Kankyū no Ame (Rain That Falls on the Ninth Day after the Onset of Cold)	寒九の雨	かんきゅうのあめ	0198
Kannazuki Shigure (Brief Rain in October)	十月時雨	かんなづきしぐれ	0989
Kanrin (Rain That Arrives When Expected or at the Perfect Moment)	甘霖	かんりん	1177
Kanro no Ame (Sweet Rain / Rain That Is the Heavens' Blessing and Moistens the Trees and Plants)	甘露の雨	かんろのあめ	1260
Kanrohōu (Sweet Dew Law Rain / The Buddha Is like Rain, Pouring His Teachings Down on People)	甘露法雨	かんろほうう	1443
Kanryū (Sweet Raindrops or Another Way to Say Drizzle)	甘霤	かんりゅう	0250
Kanshoku no Ame (The Rain That Comes on the Day before Equinox)	寒食の雨	かんしょくのあめ	1255
Kanten Jiu (Welcome and Beneficial Rain during a Drought / A Helping Hand during a Time of Great Need)	旱天慈雨	かんてんじう	0313
Kanten no Jiu (Long-Awaited Rainfall That Finally Arrives)	干天の慈雨	かんてんのじう	0232
Kanzasāme (Rain That Comes with Strong Blowing Wind—Akita)		かんざさーめ	0986

Kappa no Amagoi (Folktale about a Mythical Creature Who Was Asked to Pray for Rain, Which He Did by Begging the Sky and Not Eating or Drinking Any Water and Was Successful but Died)	河童の雨乞い	かっぱのあまごい	1678
Karasaki no Yau (Karasaki's Night Rain on an Old Pine Tree)	唐崎の夜雨	からさきのやう	0019
Karasu ga Sakan ni Mizuabi o suru Toki wa Ame ga Chikai (When the Crows Bathe, Rain Is Near)	カラスが盛んに水浴びをするときは雨が近い	からすがさかんにみずあびをするときはあめがちかい	0654
Karatsuyu (Rainy Season with Little Rain / A Dry Rainy Season)	空梅雨	からつゆ	0418
Karegiku (Chrysanthemums Withered by Rain and Snow)	枯菊	かれぎく	1560
Karetsuyu (Dry Rainy Season)	涸梅雨	かれつゆ	1264
Kasa no Hi (June 11, the Day the Rainy Season Begins Is Called Umbrella Day)	傘の日	かさのひ	1853
Kasashigure (Rain That Falls on the Traveling Wanderer's Hat)	笠時雨	かさしぐれ	1261
Kassui (When the Lack of Rainfall Dries Up the Water)	渇水	かっすい	1114
Kasumi Hajimete Tanabiku (Haze Begins / February 23, the Second Season of the Twenty-four Sekki of Rain and Water)	霞始靆	かすみはじめてたなびく	1111
Katabataami (A Local, Sudden, Summer Rain Shower)		かたばたあみ	1266
Katabui (One Side Rain / Rain That Falls on One Area, While Leaving the Sky Sunny in Another)	片降い	かたぶい	1425
Kataburi (A Rainy Spell)	偏降り	かたぶり	0446

Kataburi (Raining in One Place but Right Next to It Is Not Raining)	片降	かたぶり	0653
Katashigure (Scattered Winter Rain)	片時雨	かたしぐれ	0448
Katsuraotoko Sumazu Nari Keri Ame no Tsuki ("The Man in the Moon Is Made Homeless by the Rainy Night Moon" —Matsuo Bashō)	桂男 すまず なりけり 雨の月 （芭蕉）	かつらおとこ すまず なりけり あめのつき （ばしょう）	1440
Katsuu (Thirsting for Rain / Drought Caused by Extended Sunny Weather)	渇雨	かつう	1115
Kau (A Beautiful and Favorable Rain)	佳雨	かう	1262
Kau (A Celebratory Rain That Falls When It Ought to Fall)	嘉雨	かう	1410
Kau (A Little Rainfall)	寡雨	かう	1406
Kau (Flower Rain / Rain during Flower Blossoming Season)	華雨	かう	1263
Kau (It Is Raining / Spring Rain)	華雨	かう	0469
Kau (Rain Which Falls for a While and Stops Immediately)	過雨	かう	1417
Kau (Summer Rain)	夏雨	かう	1407
Kau (To Rain)	下雨	かう	1403
Kaun'u (Short and Brief Rain of the Summer)	過雲雨	かうんう	1408
Kawadome (River-Stopping Rain / A Heavy Summer Rain That Closes Bridges and Bans Travel over Rivers)	川止め	かわどめ	1854
Kawato no Shigure (River Sound Rain)	川音の時雨	かわとのしぐれ	1768

Kaze no Sobae (Wind That Plays / A Brief and Playful Rain with Wind)	風の戯え	かぜのそばえ	1411
Kazebana (Wind Flowers / Winter Wind Which Brings Scattered Raindrops and Snow-flakes That Fall like Flower Petals)	風花	かぜばな	0142
Kazehana (Flurry of Snow and Rain That Looks like Petals Falling / Wind Flowers)	風花	かぜはな	0165
Kazekuso (Rain That Arrives Just before the Wind Disappears, as though It Were Something the Wind Left Behind)	風くそ	かぜくそ	1265
Kazenomi (Fruit of the Wind / A Brief Rain That Comes with Wind—Aichi and Kagawa)	風の実	かぜのみ	1404
Keame (Fine Rain That Looks like a Fiber)	毛雨	けあめ	0021
Kegare (Ritual Pollution of Sacred Waters with Bones or Other Unclean Objects to Anger and Disturb the Deities into Producing Rainfall)	穢れ	けがれ	1909
Keibon no Taiu (A Rain So Heavy, It Looks like a Large Bon [Bowl] Was Turned Over)	傾盆大雨	けいぼんのたいう	0143
Keichitsu no Ame (Rain in Early Spring, Where Hibernating Animals Come Out and Rain Melts the Snow)	啓蟄の雨	けいちつのあめ	0925
Keiryou no Ame (Dragon Rain / Storm with Heavy Wind and Rain, as if a Dragon Was Twisting Itself in All Directions)	挂竜の雨	けいりょうのあめ	0941

Keiu (A Little Bit of Light Welcome Rain in Spring)	軽雨	けいう	0383
Keiu (Light Rain / Spring Rain)	軽雨	けいう	0269
Keiu (Rain in the Valleys)	渓雨	けいう	1423
Kemuriame (Smoke-like Rain)	煙雨	けむりあめ	0068
Ken'u (Sudden Rain)	懸雨	けんう	1165
Kengen (The Sky, the Source of Rain, Where the Cycle of Water Begins and Ends)	乾元	けんげん	1574
Ketsuu (Rain Mixed with Volcanic Ash)	血雨	けつう	0449
Ki no Memoyashi (Rain in Early Spring That Brings Plant Buds to Emerge from the Soil)	木の芽萌やし	きのめもやし	1419
Ki no Menagashi (Rain That Falls on Sprouting Tree Buds)	木の芽流し	きのめながし	1622
Ki no Meokoshi (Rain That Wakes the Tree Buds)	木の芽おこし	きのめおこし	1731
Kibūriami (Tiny Droplets of Rain That Fall with No Sound— Kuroshima, Okinawa)		きーぶーりあみ	1451
Kibusāruami (Raindrops That Are Fine as Mist)		きぶさーるあみ	1421
Kibutsume (Rain in the First Month of the Lunar Calendar)	木萌雨	きぶつめ	1436
Kichigaiame (Mad Rain / Just When You Think It Will Clear, It Rains Again like Madness)	気違い雨	きちがいあめ	1438
Kigetsuu (Rain That Continues through the Month and Extends to the Following Month)	騎月雨	きげつう	1732
Kiggeame (Very Heavy Rain)		きっげあめ	1862

Kin'u (A Celebratory Rain after a Long Drought)	錦雨	きんう	1733
Kinnu (Suffer from a Long Rain)	窘雨	きんう	1466
Kinu no Ame (Rain That Looks like Thin Threads of Silk Falling from the Sky)	絹の雨	きぬのあめ	1164
Kiri no Ka (Fragrance of Mist)	霧の香	きりのか	1319
Kiri no Shōben (Foggy, Mist-like Rain—Ina Region, Nagano)	霧の小便	きりのしょうべん	1321
Kiriniji (Fog Rainbow)	霧虹	きりにじ	1948
Kirisame (Fine Rain like Fog and Mist)	霧雨	きりさめ	0377
Kirishigure (Rain That Looks like Fog)	霧時雨	きりしぐれ	1803
Kirishigure Fuji o Minu Hi Zo Omoshiroki ("Mount Fuji Obscured by Foggy Rain Is More Interesting"—Matsuo Bashō)	霧時雨 富士を見ぬ日ぞ 面白き（芭蕉）	きりしぐれ ふじをみぬひぞ おもしろき（ばしょう）	1861
Kirishizuku (Droplets Made in Thick Mist)	霧雫	きりしずく	1524
Kisame (Raindrops That Fall Off the Leaves and Branches of Trees)	樹雨	きさめ	0158
Kishiu (Rain-Stopping Prayer)	祈止雨	きしう	1802
Kitaarashi (Cold and Rainy Day during Planting Season—Inba Region, Chiba)		きたあらし	1573
Kitaburi (Rain That Comes with a Cold Northern Wind in Late Autumn to Early Winter; Winter Is Coming Soon)	北降り	きたぶり	1730
Kitashibuki (A Windy Rain of the Winter That Comes from the North)	北しぶき	きたしぶき	1434

Kitashigure (Light and Soft Rain from the North)	北時雨	きたしぐれ	0115
Kitayama Shigure (Rain Coming Down from the North Mountain Kitayama Area of Kyoto)	北山時雨	きたやましぐれ	0157
Kitsune no Ame (Fox Rain / Rain That Seemingly Comes from Nowhere as if a Trick by a Fox Spirit)	狐の雨	きつねのあめ	0148
Kitsune no Goshūgi (Fox Celebration / Raining While Sunny, the Weather in Which the Foxes Have Wedding Ceremonies)	狐の御祝儀	きつねのごしゅうぎ	1852
Kitsune no Goshūgi ga Hajimaru to Kanarazu ni San Nichigo ni wa Ame ga Furu (When the Foxes Begin Their Celebration, Three Days Later It Is Guaranteed to Rain)	狐の御祝儀が始まると必ず二、三日後には雨が降る	きつねのごしゅうぎがはじまるとかならずに、さんにちごにはあめがふる	0491
Kitsune no Yomeiri (Rain That Falls Even Though the Sun Is Shining / The Day That Foxes Have Their Wedding Ceremony)	狐の嫁入り	きつねのよめいり	0126
Kitsune no Yometori Ame (Fox Taking a Wife Rain / Light Rain in Sunshine)	狐の嫁取り雨	きつねのよめとりあめ	0114
Kitsuneame (Fox Rain / Light Rain on a Sunny Day)	狐雨	きつねあめ	1531
Kitsunebiyori (Fox Weather / Raining and Sunny)	狐日和	きつねびより	1513
Kiu (A Blessed Rain That Falls after a Long Period of Sunshine)	喜雨	きう	0226
Kiu (Extraordinary, Extremely Heavy Rain That Makes One Wonder If It May Be the Work of a Demon)	鬼雨	きう	1590

Kiu (To Invoke or Pray for Rain)	祈雨	きう	1935
Kiu Itaru (Long-Awaited Rain Finally Arrives amid a Drought)	喜雨到る	きういたる	0992
Kiu Nikki (Compilation or Daily Record of Rain Prayer)	祈雨日記	きうにっき	1942
Kiufuda (Rain-Praying Tablet)	祈雨札	きうふだ	1818
Kiuhō (Rainmaking Prayer / Praying for Rain)	祈雨法	きうほう	0394
Kiuhō Nikki (Diary of Rainmaking Rituals and Prayers)	祈雨法日記	きうほうにっき	1662
Kiuki (Record of Rain Prayer)	祈雨記	きうき	1804
Kiukyō (Sutra Chanted While Praying for Rain)	祈雨経	きうきょう	1856
Kiuodori (Rain-Praying Dance)	祈雨踊り	きうおどり	1595
Kiusai (Rain-Praying Festival)	祈雨祭	きうさい	0612
Kiuyasumi (Blessed Rain Holiday / When It Finally Rained after a Drought, Farmers Would Rest and Offer Prayers of Gratitude to the Deities)	喜雨休み	きうやすみ	1901
Kizuyu (Rain That Falls When the Plum Blossoms Turn Yellow / Summer)	黄梅雨	きづゆ	1588
Koame (A Tiny Little Light Rain)	小雨	こあめ	0201
Koboreame (Rain That Spills Over from Clouds)	こぼれ雨	こぼれあめ	0347
Koburi (It Rains a Little Bit and Is Weak)	小降り	こぶり	0028
Kochi ga Hukeba Ame (When the East Wind Blows, Rain)	東風が吹けば雨	こちがふけばあめ	0003
Kodomo Sawageba Ame ga Furu (If the Children Play Loudly Outside, It Will Rain)	子供騒げば雨が降る	こどもさわげばあめがふる	1619

Kogane no Ame (Gold Coins from Heaven Rain / A Rain of Blessing after a Long Drought like Gold Falling from the Sky)	黄金の雨	こがねのあめ	0993
Kōjakuu (Yellow Sparrow Rain / Rain That Falls in the Fifth and Ninth Lunar Months)	黄雀雨	こうじゃくう	1428
Kokoro no Ame (Rain of the Heart and Spirit / When the Heart Is Not Radiant and the Mind Is Not Clear)	心の雨	こころのあめ	1189
Kokoro Shiguru (Drizzling Rain in My Heart and Soul)	心時雨る	こころしぐる	0330
Kokufū Hakuu (Black Wind / White Rain / Rain and Dust)	黒風白雨	こくふうはくう	0337
Kokun Hakuu (When the Sky Turns Dark and It Rains Heavily for a Period of Time)	黒雲白雨	こくうんはくう	0091
Kokuu (Black Rain Dark with Soot)	黒雨	こくう	0118
Kokuu (Grain Rain, a Season That Begins on April 19, from the Twenty-four Sekki Ancient Calendar)	穀雨	こくう	0403
Kokuu (Rain in the Valley)	谷雨	こくう	1530
Komaame (Fine, Thin Rain)	細雨	こまあめ	1431
Komaburu (Light Rain—Ehime)		こまぶる	0991
Kōmō (Fine Rain with Tiny Raindrops)	涳濛	こうもう	1422
Kōmoku (Sweet and Blessed Rain That Helps the Crops Grow)	膏霂	こうもく	1617
Konburi (When the Rain Becomes Weak—Aomori)		こんぶり	1432
Konkon (Much Rain / Copious Rain)		こんこん	0990

Konohashigure (Leaf Rain / Scattering and Fluttering Leaves in Winter That Sound like Rain)	木の葉時雨	このはしぐれ	1426
Konomi no Ame (Tree Nut Rain / Nuts That Fall from Trees in Autumn, Sounding like Drizzle)	木の実の雨	このみのあめ	1391
Konomi no Shigure (Nuts Falling Off Trees in Autumn That Sound like Drizzling Rain)	木の実の時雨	このみのしぐれ	0202
Konukaame (Fine Rain)	小糠雨	こぬかあめ	1764
Konukaame (Small Powder Rice Bran Rain / A Delicate Rain That Falls Lightly in Early Spring)	粉糠雨	こぬかあめ	1424
Kōriarare (Ice Hail Rain / Rain That Freezes and Falls as Ice on Snowflakes)	氷霰	こおりあられ	0645
Kōrigiri (Ice Fog / Frozen, Suspended, Fine Ice Water Drops)	氷霧	こおりぎり	0277
Kōrin (Long Rain That Nourishes the Farmlands)	膏霖	こうりん	1429
Kōrin (Long Rainfall That Lasts an Extended Period of Time)	洪霖	こうりん	1807
Kōrō (Rainwater That Accumulates on the Ground)	潢潦	こうろう	1430
Kōrō (Rainwater That Has Accumulated on the Ground)	行潦	こうろう	0928
Korodoame (Rain in a Very Narrow and Small Area—Iwate)		ころどあめ	1427
Kōryō un'u (Dragon Cloud Rain / Rise to Heaven, Encounter a Rain Cloud and Become a Dragon)	蛟竜雲雨	こうりょううんう	0396
Kosame (Light and Fine Rain / Just a Little Bit of Rain)	小雨	こさめ	0092

Kosame ga Paratsuku (Light Rain Flutters)	小雨がぱらつく	こさめがぱらつく	1442
Kosame Tokidoki Furu (Light Rains Sometimes Fall, Season Fifty-three in Seventy-two Kō Ancient Calendar, October 28– November 1)	霎時施	こさめときどきふる	1742
Kōshi no Ame (Rain on the Day of Wood Rat)	甲子の雨	こうしのあめ	1625
Koshiame (Long Rain That Falls Softly and Continues to Fall Forever)	こし雨	こしあめ	1190
Kōsui (Rainfall)	降水	こうすい	0364
Kōsui Kakuritsu (Chance of Rain)	降水確率	こうすいかくりつ	1830
Kōsuiryō (The Amount of Water and Ice That Has Fallen from the Sky to the Earth)	降水量	こうすいりょう	0919
Kōu (Pouring Rain That Continues Falling)	恒雨	こうう	1593
Kōu (Rain of Blessing / Sweet Rain / Rain That Helps the Crops Grow)	膏雨	こうう	1626
Kōu (Rain That Comes Exactly When You Were Waiting for It)	好雨	こうう	1808
Kōu (Rain That Falls on the Surface of a River)	江雨	こうう	1851
Kōu (Rain with a Pleasant Fragrance)	香雨	こうう	1537
Kōu (Rainfall)	降雨	こうう	1699
Kōu (Spring Rain Pouring Down on Flowers / Rain That Moistens the Crops)	紅雨	こうう	0365
Kōu (The Rain That Falls)	行雨	こうう	1859
Kōu (To Start Raining)	興雨	こうう	1809

Kōu (Very Heavy Rain That Could Cause a Flood)	江雨	こうう	1104
Kōu chōun (Rain and Morning Clouds—Parable about the Fellowship of a Man and a Woman)	行雨朝雲	こううちょううん	1943
Koubai no Ame (Rain That Falls When the Plum Trees Ripen and Turn Yellow)	黄梅の雨	こうばいのあめ	1858
Kougetsuu (Rain That Goes On for Months)	交月雨	こうげつう	1435
Kōrin (Rain Forest)	降雨林	こううりん	1878
Kōuryō (The Amount of Rainfall)	降雨量	こううりょう	1903
Kōuten (Rough Sky / Stormy Weather)	荒天	こうてん	1521
Kōya no Ogusonagashi (A Rain That Purifies and Cleanses the Spirit on March 22 of the Lunar Calendar)	高野の御糞流し	こうやのおぐそ ながし	1532
Kōya no Ogusonagashi (Rain That Falls in the Middle Day of the Equinoctial Week and Cleanses the World for Higan— Minamiyamato-gun, Nara Prefecture)	高野のお糞流し	こうやのおくそ ながし	0326
Koyami (Small Stop in Rain, Rain Pauses for a Little While)	小止み	こやみ	0244
Koyuki (Small Snow / Light Rain)	小雪	こゆき	0927
Kū (Heavy Rain)	苦雨	くう	0515
Kūchūki (Monster of the Air Rain / Acidic Rain)	空中鬼	くうちゅうき	0776
Kukaruamāmi (Least Salty Rain of the Year—Okinawa)		くかるあまーみ	1615
Kūmō (A Dimly Lit Scene with Light Rain and Fog)	空濛	くうもう	1871

Kumo ni Shiru (Rainmaking Ritual)	雲に汁	くもにしる	1810
Kuranotama No Mikoto (Deity Enshrined at Mimegurijinja Who Brought Rain to Farmers)		くらのたまのみこと	1274
Kuraokami (Japanese Dragon and Shinto Deity of Rain and Snow)	闇龗	くらおかみ	0415
Kurayamatsumi (High Rain Dragon—Deity of Rain)	闇山祇	くらやまつみ	1596
Kurenai no Ame (Crimson Rain / Rain Falling on Red Flowers Such as Azaleas and Pink Flowers Such as Peach and Apricot)	紅の雨	くれないのあめ	1591
Kuri no Hanarin'nu (Chestnut Flower Rain)	栗の花霖雨	くりのはなりんう	1538
Kurohae (Southerly Wind Blowing at the Start of the Rainy Season)	黒南風	くろはえ	0152
Kuroi Ame (A Black Rain That Fell in Hiroshima Immediately after the Atomic Bomb)	黒い雨	くろいあめ	0516
Kuromi (Rainy Season—Kagawa)		くろみ	1495
Kusa no Ame (Grass Rain / Rain That Falls on New Leaves and Blades of Grass That Have Sprung Up in Spring)	草の雨	くさのあめ	0481
Kusaridoi (Rain Chain)	鎖樋	くさりどい	1937
Kushi ga Tōri Nikui Toki wa Ame (When It Is Difficult to Comb, It Will Rain)	櫛が通りにくい 時は雨	くしがとおりにくい ときはあめ	0067
Kusuri Furu Sora Yo Totemoni Kanenaraba ("The Fifth Day of the Fifth Lunar Month Rain Is Called Medicine Rain, the Sky Is Filled with Medicine and Gold" —Kobayashi Issa)	薬降る 空よとて もに 金ならば （小林一茶）	くすりふる そらよ とてもに かねならば （こばやしいっさ）	1259

Kusurifuru (Medicine Rain / When It Rains around Noon on the Fifth Day of the Fifth Lunar Month, This Rainwater Is Believed to Make Effective Medicines)	薬降る	くすりふる	1105
Kuu Seifū (Bitter Rain and Terrible Wind / Strong, Cold Wind and Long-Lasting Rain)	苦雨凄風	くうせいふう	1945
Kyōfūu (Rain with a Strong, Blowing Wind)	強風雨	きょうふうう	1529
Kyōkau (Apricot Flower Rain / Gentle Spring Rain That Falls on April 5)	杏花雨	きょうかう	1460
Kyokuchiteki Ōame (Sudden and Very Strong Rain in a Narrow or Local Area)	局地的大雨	きょくちてきおおあめ	1528
Kyōu (Rain That Falls in the Canyon Valley)	峡雨	きょうう	1420
Kyōu (Raining Heavily)	強雨	きょうう	1895
Kyūu (A Sudden Rain Shower)	急雨	きゅうう	1106
Kyūu (Persistent, Long Rain)	久雨	きゅうう	1894
Kyūu (Praying for Rain to Fall)	求雨	きゅうう	0009
Kyūu Kon'u (Old Rain and Rain Now / Friend from the Past)	旧雨今雨	きゅううこんう	1896
Mabaraame (Sporadic Rain)	まばら雨	まばらあめ	0944
Madowasareru Ame (Mysterious and Deceiving Rain, Possibly the Doing of a Ghosting Fox)	惑わされる雨	まどわされるあめ	0307
Maezuyu (Before the Rainy Season Begins)	前梅雨	まえずゆ	1811
Manjyōfū'u (An Entire Town Is Hit with Rain and Wind)	満城風雨	まんじょうふうう	0259
Masaka no Toki (Just in Case / In a Time of Need / In Case of Rain)	まさかの時	まさかのとき	0943

Massuguna Ame (Straight Rain / Rain without Wind, Falling Down in Lines)	まっ直ぐな雨	まっすぐなあめ	0942
Matomefuru (Heavy Rains, as if Ten Days or One Month's Worth of Rain Fell at Once)	まとめ降る	まとめふる	1938
Matōu (Sword-Sharpening Rain on May 13—The Rain That Sharpened Guan Yu's Sword as He Crossed the Yangtze River)	磨刀雨	まとうう	1936
Matsukaze no Shigure (Sound of Drizzle in the Pines)	松風の時雨	まつかぜのしぐれ	0094
Matsuu (The Overflowing of a Rainwater Puddle after a Heavy Rain Shower)	沫雨	まつう	1812
Megi no Ame (Rain on Tree Buds in Spring)	芽木の雨	めぎのあめ	1897
Megumi no Ame (Welcome and Blessed Rain after a Long, Dry Period)	恵の雨	めぐみのあめ	0283
Megurushigure (Rain and Clouds Sent by the Wind toward the Mountains)	めぐる時雨	めぐるしぐれ	1657
Mei (Light Rain That Makes the Sky Dark)	溟	めい	0550
Meidou (Rain That Falls with Thunder and Lightning)	瞑怒雨	めいどう	1946
Meiu (Rain Falling from a Darkened Sky)	溟雨	めいう	1656
Menmen to (Uninterrupted Rain)	綿々と	めんめんと	1653
Mi o Shiru Ame (Rain That Knows You / Rain That Seems Aware of Your Condition of Happiness or Unhappiness and Falls at the Perfect Time)	身を知る雨	みをしるあめ	1659

Mienuame (Invisible Rain / Rain So Delicate You Can Hardly See It)	見えぬ雨	みえぬあめ	1872
Mijikayo no Ame (Rain of the Short Night / Rain on Summer Equinox)	短夜の雨	みじかよのあめ	1651
Mikumari-Kami (A Group of Water-Dividing Deities Who Are the Object of Worship and Prayed to in Rites and Rituals Invoking Rain)	水分神	みくまりかみ	1880
Mimeguri Jinja (Shrine for Kuranotama-No-Mikoto Who Brought Rain to Farmers Suffering from Drought)	三囲神社	みめぐりじんじゃ	0782
Mino (Straw Raincoat)	蓑	みの	0053
Mino ni Nari Kasa ni Nari (Protecting from the Rain, Protecting from the Sun)	蓑になり傘になり	みのになりかさになり	1813
Minokasa (A Straw Rain Cape)	蓑笠	みのかさ	0149
Mishigeru (Mountainous Dialect for Rain—Nishiokitama, Yamagata)		みしげる	0914
Mitsu Unfūu (There Are Clouds in the Sky but There Is No Rain / When Things in Life Do Not Have the Outcome One Expected)	密雲不雨	みつうんふう	1068
Mitsuhanome no Kami (Rain and Water Goddess)	弥都波能売神	みつはのめのかみ	1815
Mitsuka Ame (After the Thunderstorm, to Have Time to Bind Three Bundles of Cut Rice Plants before the Next Rain Starts)	三束雨	みつかあめ	0428
Mitsuu (Honey Rain / Light Rain)	蜜雨	みつう	0174

Miuchūu Byū (Prepared for a Future Disaster / Birds Repairing Their Nests before a Large Rain)	未雨綢繆	みうちゅうびゅう	1814
Mizore (Rain Mixed with Snow Sleet)	霙	みぞれ	0433
Mizore Majiri No Ame (Rain Mixed with Hail)	みぞれ混じりの雨	みぞれまじりのあめ	0476
Mizore Majiri No Yuki (Sleet and Hailstone / White Opaque Ice Particles Seen Falling from the Clouds)	みぞれ混じりの雪	みぞれまじりのゆき	0167
Mizoruru (When Sleet Falls, Mixed with Rain and Snow)		みぞるる	0916
Mizuamasu (Flood from Rain)		みずあます	1515
Mizuaoi (Water Rain Flower / A Flower That Grows Wild in Rice Paddies / Summer)	雨久花	みずあおい	1273
Mizugaminari (Thunder with Rain / Lightning That Does Not Cause Fires)	水神鳴り	みずがみなり	0479
Mizugaminari (Thunder with Rain / Lightning That Does Not Cause Fires)	水雷	みずがみなり	0723
Mizuhanome-No-Mikoto (Female Rain Deity)	水波能売神	みずはのめのみこと	1546
Mizutamari (Rainwater That Collects in a Hollow in the Ground)	水溜	みずたまり	1816
Mizutoriame (Water-Taking Rain / May Rain That Collects for Rice Planting)	水取雨	みずとりあめ	0917
Mizuyuki (Water Snow / Sleet That Falls)	水雪	みずゆき	0923
Modorizuyu (Rain That Is Reminiscent of the Rainy Season / To Return to the Rainy Season)	戻り梅雨	もどりづゆ	0915

Mogikurai (Rain That Falls in Early Summer and Destroys the Wheat Harvest)	麦食らい	もぎくらい	1178
Mokuu (To Be Soaked by Rain)	沐雨	もくう	1947
Mokuu shippū (Working Hard While Being Exposed to Rain and Wind / Facing Various Hardships)	沐雨櫛風	もくうしっぷう	0150
Moratta Ami (Received Rain / Sudden Rain—Okinawa)		もらったあみ	1069
Mōu (A Light Rain Falls, Darkening the Sky)	濛雨	もうう	0830
Mōu (Rain So Intense You Cannot See Anything around You)	盲雨	もうう	0918
Mōu (Raining Very Hard)	猛雨	もうう	1385
Mouretsuna Ame (Furious Rain / Very Heavy Rain That Could Potentially Cause a Disaster)	猛烈な雨	もうれつなあめ	1238
Moya (Falling Rain or Snow)	霏	もや	0138
Mugetsu (The Mid-Autumn Full Moon Is Made Invisible by Rain)	無月	むげつ	1383
Mugiame (Wheat Rain That Is Very Fine and Very Weak)	麦雨	むぎあめ	0054
Mugiarashi (Storm on the Barley Fields)	麦嵐	むぎあらし	1956
Mukaezuyu (Rain That Lasts Several Days before the Rainy Season and Ushers in the Rainy Season)	迎え梅雨	むかえづゆ	0022
Murakumogō (A Katana Called Cloud Pile River Because When Hideyoshi Saw the Sword, He Said It Looked like a Raining Cloud)	村雲江	むらくもごう	1944

Murasame (Passing Rain / It Pours Heavily and Stops Immediately)	群雨	むらさめ	0257
Murasame (Rain on the Village / A Rain That Makes an Especially Pleasant Sound on Thatched Roofs of a Village)	村雨	むらさめ	0523
Murasame [Katana] (Village Rain—Japanese Sword Struck by Tsuda Echizen No Mori Sukehiro from Nanban Iron)	村雨	むらさめ	1394
Murasame Matsukaze (Village Rain and Pine Wind, Two Sister Ama Divers Who Became Nuns in the Heian Period)	村雨・松風	むらさめ・まつかぜ	1388
Murashigure (Late Autumn or Early Winter Rain That Falls Briefly and Then Stops)	群時雨	むらしぐれ	1904
Murashigure (Village Rain / Heavy Rain That Passes By Immediately / The Sound of Rain on the Thatched Roofs of the Village)	村時雨	むらしぐれ	0436
Murashigure (Village Rain / Short Rain That Suddenly Stops)	叢時雨	むらしぐれ	1660
Murōyama (Mountain Where a Rain Dragon Deity Lives and Has Been Used in Rain Prayer Ceremonies since Ancient Times, Nara)	室生山	むろうやま	1870
Mushishigure (Insect Rain / The Voices of Many Insects in Autumn Crying Together Sounds like Drizzling Rain)	虫時雨	むししぐれ	1985
Muu (Very Fine Rain That Feels like a Mist)	霧雨	むう	1239
Nabewari (Early Autumn Long Rain)	鍋割	なべわり	1669

Nagaame (A Long Spell of Rain That Lasts for Several Days)	長雨	ながあめ	0306
Nagaami (Rain Season / A Long Spell of Rain—Okinawa)	ナガアミ	ながあみ	1516
Nagame (A Long Rain Lasting Several Days)	霖	ながめ	1668
Nagameimi (Abstention during the Long Rain / May)	霖雨斎み	ながめいみ	1958
Nagase (Rainy Season—Tokushima)	長雨	ながせ	1905
Nagashi (Rain That Sheds Things and Falls for a Long Time / Rainy Season—Kagoshima)	流し	ながし	1973
Nagashi (The Rain Season—Kyushu)	流し	ながし	1906
Nagashi (To Be Swept Away)	流し	ながし	1275
Nagashiame (Rain That Washes or Flows Away)	流し雨	ながしあめ	1661
Nagashihae (The Rainy Season—Kyushu)	ながし南風	ながしはえ	1910
Nagashita (The Rain Season—Kagoshima)		ながした	1995
Nagatsuki no Shigure (Rain of October)	九月の時雨	ながつきのしぐれ	1667
Nagatsuyu (When the Rain Season Is Long and Continuous)	長梅雨	ながつゆ	1666
Nago no Shōben (Mist and Fog-like Rain—Hamana Region, Shizuoka)	なごの小便	なごのしょうべん	1237
Nagurisame (Sideways Rain / Rain and a Very Strong Side Wind)		なぐりさめ	0665

Nakabiyori (Amid a Long Rain, the Weather Slightly Changes)	中日和	なかびより	0664
Nakaburi (Rain on Summer Equinox)	中降り	なかぶり	0660
Nakidashisōna Soramoyō (The Sky Appears as though It Is about to Start Crying)	泣き出しそうな空模様	なきだしそうなそらもよう	0520
Naku Namida Ame to Furanamu Watarigawa Mizu Masarinaba Kaeri Kurugani ("I Want My Tears to Fall as Rain. Then the Waters of the Sanzu River Would Rise, and She Would Give Up Crossing It and Come Back to Me."—Ono No Takamura, Heian poet)	泣く涙 雨と降らなむ 渡り川水増さりなば 帰り来るがに(小野篁)	なくなみだ あめとふらなむ わたりがわ みずまさりなば かえりくるがに(おののたかむら)	1950
Namiame (Normal and Ordinary Rain / Average Rain)	並雨	なみあめ	0662
Namida no Shigure (Rain of Tears)	泪の時雨	なみだのしぐれ	1767
Namida no Shigure (Rain That Falls like Tears)	涙の時雨	なみだのしぐれ	0416
Namidanoame (Rain of Tears)	涙の雨	なみだのあめ	1634
Naname Ni Futtekuru Ame (Sloping Rain That Falls Diagonally)	斜めに降ってくる雨	ななめにふってくるあめ	0892
Nanatsusagari no Ame (Rain That Starts at 4 PM and Continues for a Long Time and Will Not Stop)	七つ下がりの雨	ななつさがりのあめ	0666
Nanzan ni Kumo o Okoshi Hokuzan ni Ame o Kudasu ("Clouds in the Southern Mountains, Rain in the Northern Mountains"—Zengo)	南山起雲北山下雨	なんざんにくもをおこしほくざんにあめをくだす	1873
Naramairi (When It Is about to Rain Soon in Nara)	奈良参	ならまいり	1117

Narukami no Sukoshi Toyomite Furazutomo Wawa Todoma Ramu Imo-shi Todo Meba ("Faint Sound of Thunder, Even If It Does Not Rain, I Will Stay Here with You"—Man'yōshū [Collection of Ten Thousand Leaves], Volume 11, Song 2514)	鳴る神の 少し響みて 降らずとも 我は留まらむ 妹し留めば(万葉集 第十一巻 二千五百十四番歌)	なるかみの すこし とよみて ふらず ともわはとどまらむ いもしとどめば (まんようしゅう だい11かん 2514 ばんか)	1734
Narukami no Sukoshi Toyomite Sashikumori Ame mo Furanuka Kimi o Todomemu ("Faint Clap of Thunder, Cloudy Skies, It May Rain and If So, Will You Stay Together with Me"—Man'yōshū [Collection of Ten Thousand Leaves]. Volume 11, Song 2513)	鳴る神の 少し響みて さし曇り雨も降らぬか 君を留めむ(万葉集 第十一巻 二千五百十三番歌)	なるかみの すこしと よみて さしくもりあ めもふらぬか きみ をとどめむ(まんよう しゅう だい11かん 2513ばんか)	1737
Natanetsuyu (It Rains Gently from March to April When the Nano-hana [Rapeseed] Flowers Are in Bloom, Early Spring)	菜種梅雨	なたねつゆ	0522
Natsu ro Ame (Summer Rain)	夏の雨	なつのあめ	1116
Natsu ro Chō (Summer Butterflies, If They Are Flying Low, Rain Will Soon Arrive)	夏の蝶	なつのちょう	1549
Natsugure (Summer Evening Rainstorm)	夏ぐれ	なつぐれ	0661
Natsusame (Summer Rain)	夏雨	なつさめ	1302
Natsushigure (Summer Rain That Falls and Stops Intermittently)	夏時雨	なつしぐれ	1118
Neko ga Kao o Arau to Ame (When the Cat Washes His Face, It Will Rain)	猫が顔を洗うと雨	ねこがかおをあらう とあめ	0485
Nekogeame (Cat Hair Rain That Is Disliked by the Wheat Farmers in Fukuoka)	猫毛雨	ねこげあめ	1168
Nekomakuri (Rolling Cats / Rain-water Flowing Violently Downhill)	猫まくり	ねこまくり	1479

Nekonkeame (Fog-like and Misty Rain)	猫毛雨	ねこんけあめ	1301
Nenmatsuzuyu (When It Rains like the Rainy Season but in Winter, December)	年末梅雨	ねんまつづゆ	0663
Nettaiurin Kikō (Wet Forest / Rainforest Climate)	熱帯雨林気候	ねったいうりんきこう	0667
Nichinichi kore kōjitsu ("Sunny Days, Rainy Days, All Days Are Wonderful," from the Blue Cliff Record Compilation of Zen Buddhist Kōans published in 1125—Zengo)	日々是好日（碧巌録）	にちにちこれこうじつ（へきがんろく）	0545
Nichiuryō (Daily Rainfall)	日雨量	にちうりょう	1303
Niji no Shōben (When It Rains under Bright Sunlight—Tokushima)	虹の小便	にじのしょうべん	0659
Nijiriagari (When the Rain Finally Subsides and the Sun Comes Out a Little Bit)	躙上	にじりあがり	0668
Nisemono no Shigure (Fake Rain)	偽物の時雨	にせもののしぐれ	1480
Nishi Kaze wa Ame ga Ōi (The Western Winds Bringing Lots of Rain)	西風は雨が多い	にしかぜはあめがおおい	0055
Nishiage (Rain That Starts and Stops)	西上	にしあげ	0996
Niwaka Biyori (A Sudden Break in the Rain and Burst of Sunshine)	俄日和	にわかびより	0407
Niwakaame (Rain That Starts Suddenly and Stops Very Soon)	俄雨	にわかあめ	0621
Niwakaame (Temporary and Transient Rain That Stops Very Soon)	にわか雨	にわかあめ	0046
Niwakayuki (Heavy and Sporadic, Transient Snow and Frozen Rain)	にわか雪	にわかゆき	0251

Niwatazumi (Heavy Rainfall)	潦	にわたずみ	0349
Niwatori ga Osoku made Soto ni Iru Toki wa Yokujitsu Ame ni Naru (When the Chickens Stay Out Late, It Will Rain the Next Day)	鶏が遅くまで外にいるときは、翌日雨になる	にわとりがおそくまでそとにいるときは、よくじつあめになる	0620
Nochi no Murasame (Sudden Rain Shower in Autumn / Autumn Village Rain)	後の村雨	のちのむらさめ	0619
Nōin (Heian-Era Monk and Poet Widely Known for His Poems That Served as Rainmaking Prayers)	能因	のういん	1837
Noki no Itomizu (Thread Water of the Eaves / Thin Raindrops like Fine Thread Falling from the Edge of Eaves)	軒の糸水	のきのいとみず	1070
Noki no Tamamizu (Raindrops Falling from Eaves)	軒の玉水	のきのたまみず	1191
Nokidoi (Rain Collected by the Roof)	軒樋	のきどい	1271
Nokoritsuyu (Long Rain That Falls Again after the Rainy Season Has Ended / Rain That Is Left Behind)	残り梅雨	のこりつゆ	0997
Nokorizuyu (When It Rains Even Though the Rainy Season Has Already Passed / Rain That Is Left Behind)	残り梅雨	のこりづゆ	0618
Nōmu (Dense Mist and Heavy Fog-like Rain)	濃霧	のうむ	0062
Nori no Ame (The Law of Rain / Rain That Nourishes All)	法の雨	のりのあめ	1961
Nozokiame (Peeping Rain / When It Unpredictably Rains and Then Stops—Yamanashi)	覗雨	のぞきあめ	0994

Nozomuru wa Onaji Kumoi o Ikanareba Obotsukanasa o Sofuru Shigure Zo ("We Are Looking at the Same Sky, Why Does the Rain Make Me Miss You So Much?"—Murasaki Shikibu, *The Tale of Genji*, Soukaku, Chapter 47)	眺むるは同じ雲居をいかなればおぼつかなさを添ふる時雨ぞ（源氏物語 第四十七帖 総角）	のぞむるはおなじくもいをいかなればおぼつかなさをそふるしぐれぞ（げんじものがたり だい47ちょう そうかく）	1572
Nukaame (Very Fine Rain That Falls in Spring)	糠雨	ぬかあめ	0831
Nukeburi (When It Drizzles as if the Bottom of a Heavenly Bucket Fell Off)	抜降	ぬけぶり	0995
Nureenn (Rainwater Edge / Veranda with Rain Protection)	濡れ縁	ぬれえん	1951
Nureginu (Rain-Spattered Robe / Wet Clothing)	濡衣	ぬれぎぬ	1952
Nurenu Saki no Kasa (Umbrella before Rain—Parable for Preparing in Advance so as Not to Fail)	濡れぬ先の傘	ぬれぬさきのかさ	0458
Nurenuame (The Wind Blowing through the Pine Trees, the Sound Is Very Like Rain Falling)	濡れぬ雨	ぬれぬあめ	1502
Nureonna (Rain That Drenches a Woman / A Type of Ghost or Monster)	濡女	ぬれおんな	0738
Nuresobotsu (Being Drenched from the Rain)	濡れそぼつ	ぬれそぼつ	0998
Nyūbai (Rain That Signals the Rainy Season Begins, Entering the Rainy Season)	入梅	にゅうばい	0056
Nyūdōgumo (Cloud That Brings Heavy Rain / Cumulonimbus)	入道雲	にゅうどうぐも	1997
Nyūeki (Beginning of the Winter Drizzle Season)	入液	にゅうえき	0183

Oai (Rainmaking Practice of Annoying or Upsetting Deities into Causing Rain by Defiling Sacred Water Sources with Unclean Things Such as Bones)	汚穢	おあい	1953
Ōame (Great Rain / Large Amount of Rain)	大雨	おおあめ	0521
Ōarashi (Raging Storm)	大嵐	おおあらし	0420
Ōare (Intense Weather with Heavy Rain and Blowing Wind / Storm)	大荒れ	おおあれ	0709
Ōburi (Storm with Heavy Rain)	大降り	おおぶり	1499
Ochiba no Ame (Rain of Fallen Leaves / Autumn Leaves Falling like Rain)	落葉の雨	おちばのあめ	0409
Ochiba no Shigure (Leaves Falling Continuously, Sounding like Winter Drizzling Rain)	落葉の時雨	おちばのしぐれ	1827
Odayamu (Rain Becomes Gentle)	小弛む	おだやむ	0729
Odoshiame (Threatening Rain / A Sudden Rain That Feels as though the Sky Was Intimidating You)	脅し雨	おどしあめ	0737
Oharai (Religious Purification for Ceremonies Including Rainmaking Prayer)	お祓い	おはらい	1824
Ohisanname (Rain despite Sunny Weather)	御日さん雨	おひさんあめ	1243
Ohyama afuri Jinja (Shrine atop a Rainmaking Mountain in Kanagawa—If the Top of the Mountain Is Veiled in Clouds, It Is Believed to Bring Rain Momentarily)	大山阿夫利神社	おおやまあふり じんじゃ	1834
Oikakete Arare ni Korobu Chidori Kana (Like a Plover That Chases and Falls to Hail—Haiku, Chōda)	追かけて 霰に ころぶ 千鳥かな （蔦雫）	おいかけて あられに ころぶ ちどりかな （ちょうだ）	1343

Okadatsuame (Thunderstorm—Yamagata)	おかだつ雨	おかだつあめ	1691
Okami (Deity and Dragon of Rain and Snow)	靇	おかみ	1522
Okami-No-Kami (Legendary Dragon and Deity of Rain)	淤加美神	おかみのかみ	1126
Okimuradachi (Sudden Evening Shower That Comes from the East—Chita Region, Aichi Prefecture)		おきむらだち	1192
Okisakidachi (Rain That Begins to Fall from the Sea in the Southern Part of Kochi Prefecture)	沖早立	おきさきだち	1194
Okuribaiu (Downpour to See Off the Ending of the Rainy Season)	送り梅雨	おくりばいう	0314
Okuritsuyu (Rain with Thunder That Falls at the End of the Rainy Season)	送り梅雨	おくりつゆ	0058
Okurizuyu (Last Rain of the Rainy Season, Sometimes with Thunderstorms)	送り梅雨	おくりづゆ	0495
Okusonagashi (Rain That Falls in the Middle of the Day on the Equinox—Gifu)	お糞流し	おくそながし	1272
Omizukari (Rainmaking Prayer Ritual Where Sacred Water Is Carried Home in Bare Hands)	御水借	おみずかり	0907
Omoshiroshi Yuki ni ya Naran Fuyu no Ame ("Interestingly the Snow Becomes Winter Rain" —Matsuo Bashō)	面白し 雪にや ならん 冬の雨 （芭蕉）	おもしろし ゆきにや ならん ふゆのあめ （ばしょう）	1777
Onfuri (Honorable Downpour Rain on the First Day of the Year That Blesses the Harvest / January 1)	御降り	おんふり	1206

Oniarai (Rain on New Year's Eve / Rain That Washes the Demons)	鬼洗い	おにあらい	0057
Onibi (Demon Fire or Foxfire / Will-o'-the-Wisp Seen on Rainy Nights)	鬼火	おにび	1393
Oniwaarai (Rain after a Religious Ritual)	御庭洗	おにわあらい	1482
Onnaame (Rain That Falls Slowly / Woman's Rain)	女雨	おんなあめ	0004
Onnazuyu (Woman's Rain / Long and Continuing Rain)	女梅雨	おんなづゆ	1487
Onpū (In Late Summer toward the End of the Rainy Season, Wet Air and Wind Come from the Ocean)	温風	おんぷう	0920
Ontaikau kikō (Temperate, Mild Climate and Rainy Summer)	温帯夏雨気候	おんたいかうきこう	1483
Ōnuke (Very Heavy Rain as if a Plug Has Been Pulled Out of the Sky and All the Water Has Fallen Out)	大抜け	おおぬけ	1501
Ōotsubu (Rain with Large Raindrops)	大粒の雨	おおつぶのあめ	1489
Oraisanname (Evening Shower or Sunny Rain—Miyagi)	御雷様雨	おらいさんあめ	1504
Ori Kra no Ame (Rain That Comes Just after a Drought, Just in Time)	折からの雨	おりからのあめ	1197
Ōroi (Blessed Rain—Kagawa)		おうろい	1507
Ōrui (Blessed Rain—Kagawa)		おうるい	0296
Osagari (If It Rains, the Harvest Will Be Blessed This Year / New Year's Rain)	御降り	おさがり	0519
Ōshajiku (Heavy Rain—Niigata)	大車軸	おおしゃじく	1485

Ōshike (Great Storm / Large Rainstorm at Sea)	大時化	おおしげ	1833
Oshimeri (Rain That Falls Just Enough to Moisten the Ground / Rain That Is Perfectly Appropriate for the Time of Year and Amount)	御湿り	おしめり	0922
Ōshin no ame (Rain Falling on Rice Plants)	秒針の雨	おうしんのあめ	1506
Otenkiame (Nice Weather Sunny Rain)	お天気雨	おてんきあめ	1629
Otokoburi (Man's Rain / Heavy Downpour of Rain)	男降り	おとこぶり	1583
Otokotsuyu (Man's Rain / Rain That Falls Heavily and Then Stops Suddenly)	男梅雨	おとこつゆ	1009
Ōtsubu Kotsubu (Big Drops and Little Drops, the Rain Falls)	大粒小粒	おおつぶこつぶ	0971
Oyakataame (Rain That Falls Only in the Middle of the Night—Takeno Region, Kyoto)	親方雨	おやかたあめ	0746
Oyamaarai (Washing the Mountain / The Rain That Cleanses Mount Fuji)	御山洗	おやまあらい	0224
Oyami (It Stops Raining for a While)	小止み	おやみ	1576
Ozuare (Raining despite a Clear Sky—Yamagata)		おづあれ	1835
Parapara (The Sound of Raindrops)	パラパラ	ぱらぱら	0029
Paraparaame (Spotty Rain That Sprinkles)	パラパラ雨	ぱらぱらあめ	0225
Parari (The Sound of Rain or Very Slight Objects Falling Sparsely or Faintly)		ぱらり	0444

Paratsuku (Just a Little Bit of Rain)		ぱらつく	1198
Piripiri (A Little Light Rain—Kameoka Region, Kyoto)		ぴりぴり	1011
Pota Pota (The Sound of Raindrops and Tears)		ぽたぽた	1547
Potari (The Sound of Raindrops)		ぽたり	1793
Potopoto (Sound of Rain Falling in Big Raindrops)		ぽとぽと	0497
Potsupotsu (Raindrops Falling Here and There)	ポツポツ	ぽつぽつ	0832
Potsuripotsuri (Rain Falling in Drops)		ぽつりぽつり	0252
Ragetsu (The Moon after the Rain / Refreshing and Clear Mind and Heart)	蘿月	らげつ	1536
Raiden (Thunder and Lightning)	雷電	らいでん	1314
Raijin (Deity of Lightning, Thunder, and Storms)	雷神	らいじん	1598
Raiu (Rain and Snow with Lightning)	雷雨	らいう	1276
Rakuu (Falling Drops / It Is Raining)	落雨	らくう	1010
Ran'u (Sufficient Rain / A Lot of Rain)	爛雨	らんう	0747
Ran'un (Rain Cloud)	乱雲	らんうん	1491
Ransōun (Dark Gray Cloud That Brings Rain and Snow and Completely Hides the Sun and Moon)	乱層雲	らんそううん	1954
Rei (Spirit / Pray to God to Make It Rain)	靈	れい	1195
Reirin (Spirit Rain / A Mysterious Ghost Rain)	霊霖	れいりん	1631

Reiu (Blessed Rain That Falls Just When People Want It)	霊雨	れいう	0783
Reiu (Cold Rain in Late Autumn)	冷雨	れいう	0498
Reiu (Rain That Falls Quietly)	零雨	れいう	0756
Reiu (Rain-Thanking)	礼雨	れいう	1982
Ren'u (Continuous Rain Falling Every Day, for Days on End)	連雨	れんう	1277
Rensen (Fine Rain)	廉繊	れんせん	0921
Rihau (Rain That Loosens the Earth for Farmers to Cultivate)	犂把雨	りはう	1503
Rika Isshi no Haru ("A Branch of Pear Flower, Spring, Rain / A Sad and Beautiful Weeping Woman's Face Looks like a Pear Flower Wet with Rain"—Zengo)	梨花一枝春	りかいっしのはる	0121
Rikki (Yin and Yang, Rain, Wind, Light, and Dark)	六気	りっき	1279
Rikyūnezumi no Ame (Mouse-Colored, Gray Rain That Is Faint and Fine, Falling on the Beach)	利休鼠の雨	りきゅうねずみの あめ	1575
Rin (Long-Lasting Rain)	霖	りん	1484
Rin'u (A Long Rain That Continues for Days)	霖雨	りんう	1122
Rin'u (Downpour)	淋雨	りんう	0682
Rin'u Sousei (Long and Welcoming Rain That Gives Many Blessings to People and Makes Vegetation Grow)	霖雨蒼生	りんうそうせい	1085
Rin'yo (After the Rain)	霖余	りんよ	0572
Rindō (A Long Rain That Makes the Road Slippery and Muddy)	霖淖	りんどう	1124
Ringaiu (Rain Falling outside the Forest)	林外雨	りんがいう	1836

Rinju (Long, Continuous Rain)	霖澍	りんじゅ	0573
Rinnaiu (The Rain That Falls inside a Forest, Rainwater on Leaves and Running down Tree Trunks)	林内雨	りんないう	1318
Rinreki (Rain That Will Not Stop)	霖瀝	りんれき	1123
Rinrin (Continuous Rain for a Long Period of Time)	霖々	りんりん	0480
Rinrin (Rain That Does Not Stop)	霖々	りんりん	1193
Rinrō (Cold Rain in Autumn)	惏露雨	りんろう	1127
Rinsui (Flood Caused by Extended and Heavy Rains)	霖水	りんすい	1086
Rira no Ame (Rain That Falls on Lilac Flowers in a Cold Region near the Sea)	リラの雨	りらのあめ	0570
Risshun no Ame (Rain on the Spring Equinox)	立春の雨	りっしゅんのあめ	0571
Rō (Heavy Rainfall)	潦	ろう	0437
Rojison (Collecting Rainwater)	路地尊	ろじそん	1316
Rōjyu Hana Hiraku Shun'u no Mae (The Old Tree Blossoms before the Spring Rain / The State of a Buddhist Practitioner Who Has Been Diligent)	老樹花開春雨前	ろうじゅはなひらく、しゅんうのまえ	1912
Rokki (Six Elements between Heaven and Earth, Yin, Yan, Wind, Rain, Darkness, Light)	六気	ろっき	1125
Ruiu (Rain Tears That Fall during Times of Sorrow)	涙雨	るいう	0681
Ruu (A Light and Intermittent Rain like Murasame)	屢雨	るう	1071
Ruu (Intermittent Rain)	屢雨	るう	0574
Ryaku u (An Intense, Strong Rain That Almost Feels like It Is Hitting or Striking You)	掠雨	りゃくう	1179

Ryō (Rain / Water That Falls in Drops)	凌	りょう	1196
Ryō (To Keep the Rain Out)	凌	りょう	1278
Ryōin (A Rain That Falls during the Summer and Provides Coolness)	涼雨	りょういん	0144
Ryōjun (Dragon Rain / When the Dragon Cries, Rain Will Fall and Life Will Be Nourished)	竜潤	りょうじゅん	0834
Ryokuu (Early Summer Rain)	緑雨	りょくう	0465
Ryōten (Raining While the Sky Is Sunny and Clear)	両天	りょうてん	1486
Ryōu (Cool Rain)	涼雨	りょうう	0186
Ryōu (Heavy, Intensely Strong Rain)	凌雨	りょうう	1317
Ryū (Raindrops Falling from the Eaves)	霤	りゅう	0350
Ryūjinsama (Dragon God Controls the Rain and Brings the Grace of Rain to the Crops)	龍神さま	りゅうじんさま	0209
Ryūkakuu (Starting to Rain as if to Prevent My Guest from Leaving)	留客雨	りゅうかくう	1817
Ryūnen-Ji (Temple of the Dragon Rain Deity, Monks Chanted Sutras and Threw Their Kesa Robes into the Sea and Successfully Produced Rain)	龍拈寺	りゅうねんじ	1959
Ryūseiu (A Rain of Flowing, Falling Stars / Meteor Shower)	流星雨	りゅうせいう	1332
Sada' (Drizzling Rain—Kagoshima)		さだっ	1384
Sadamenaki Ame (Unpredictable and Indefinite Rain in Late Autumn to Early Winter, It Rains Suddenly and Briefly and Then Stops but Rains Again)	定めなき雨	さだめなきあめ	1282

Saikau (Rain That Makes the Flowers Bloom)	菜花雨	さいかう	1493
Saikau (Sparse and Cold Rain on the Ninth Day of the Ninth Lunar Month)	催花雨	さいかう	1280
Saikau (Spring Rain That Encourages Flowers and Hastens Their Blooms)	催禾雨	さいかう	1180
Sairuiu (Rain That Falls on Tanabata)	催涙雨	さいるいう	1362
Sairuiu (Tears Shed by Orihime and Hikoboshi When They Cannot Meet Because of the Rain / Rain on Tanabata, July 7)	酒涙雨	さいるいう	0348
Saiu (Faint Rain)	霧雨	さいう	1869
Saiu (Misty Rain)	細雨	さいう	0342
Sajikeru (Rain and Wild Wind—Oshima Region, Tokyo)		さじける	1181
Sakujitsu Ha Ame, Konnichi Ha Hare ("It Rained Yesterday but Today Is Sunny / Appreciation for Nature and Acceptance of the Natural Way of Things"—Zengo)	昨日雨今日晴	さくじつはあめ こんにちははれ	1283
Sakuraame (Rain That Falls during the Season of Cherry Blossoms)	桜雨	さくらあめ	0441
Sakuranagashi (Rain That Falls during the Season of Cherry Blossoms)	桜流	さくらながし	0359
Sakurashigure (Raining on Cherry Blossoms)	桜時雨	さくらしぐれ	1561
Sakuya No Ame (Last Night's Rain)	昨夜の雨	さくやのあめ	0767
Same (Rain)	雨	さめ	1911

Samidare (Rain of the Fifth Lunar Month / Persistent Rain That Falls in Spring or Early Summer)	五月雨	さみだれ	0366
Samidare ni Onmono Dooya Tsuki no Kao ("Rain in May Hides the Face of the Moon and Imperial Treasure"—Matsuo Bashō)	五月雨に 御物 遠や 月の顔 （芭蕉）	さみだれに おんもの どおや つきのかお （ばしょう）	1437
Samidare no Sora Da ni Sumeru Tsukikage ni Namida no Ame ha Waruruma mo Nashi ("Moon Shadow in the May Rain, My Tears Unending" —Akazome Emon)	五月雨の 空だ にすめる 月影に 涙の雨は はるる まもなし（赤染 衛門）	さみだれの そらだ にすめる つきかげ に なみだのあめは はるるまもなし （あかぞめえもん）	1405
Samidare o Atsumete Hayashi Mogamigawa ("Early Summer Rains Gathered by the Quickening Mogami River"—Matsuo Bashō)	五月雨を あつめ て早し 最上川 （芭蕉）	さみだれを あつめて はやし もがみがわ （ばしょう）	0598
Samidarebare (Brief Sunshine between Heavy Rains during the Rainy Season)	五月雨晴れ	さみだればれ	0471
Samidareboshi (May Rain Star / Star That Is Luminous Even through the Rain of the Fifth Lunar Month)	五月雨星	さみだれぼし	1413
Samidaregō (The Name of a Katana with a Hamon Blade Pattern That Looks like May Rain and Fog)	五月雨江	さみだれごう	1963
Samidaregumo (Rainclouds in the Fifth Lunar Month)	五月雨雲	さみだれぐも	0063
Samidareshiki (Intermittent and Frequent Rains in May / The Name of the Fifth Lunar Month)	五月雨式	さみだれしき	1299
Samidarezuki (Fifth Month, the Rainy Month in the Lunar Calendar)	五月雨月	さみだれづき	1361
Samidaruru (Raining in the Fifth Lunar Month)	五月雨る	さみだるる	0113

San'u (Mountain Rain / Rain That Begins to Fall from the Mountains)	山雨	さんう	0175
San'u (Rains Three Times)	山雨	さんう	1363
San'u Kitaran to Shite Kazerō ni Mitsu (If the Mountain Rains Come, the Tower Will Be Blown Over by the Wind)	山雨来たらんとして風楼に満つ	さんうきたらんとしてかぜろうにみつ	1376
San'uki (Umbrella Rain That Falls on May 6, the Anniversary of the Death of Poet Mantaro Kubota)	傘雨忌	さんうき	1962
San'un Kyōu (Clouds Floating near a Bridge in the Mountains, Rain Falling in the Valley)	桟雲峡雨	さんうんきょうう	0246
Sanbaine (A Sudden Evening Storm That Occurs So Quickly, One Has No Time to Make Even Three Bundles of Rice)	三把稲	さんばいね	1152
Sanjūrokuu (Thirty-six Rains per Year, Rain Every Ten Days, Makes for a Peaceful World)	三十六雨	さんじゅうろくう	1960
Sankayōu (Rain That Falls on the Sankayōu Flowers, Their Petals Turn from White to Transparent When it Rains)	荷葉雨	さんかようう	0070
Sansan (Rainfall / Shedding Tears)	潸潸	さんさん	1765
Sansan (The Way the Rain Falls Perpetually and Unceasingly)	潸々	さんさん	1628
Sansashigure (Onomatopoeia Rain Song—Celebratory Folk Song Sung since the Edo Period with Hand Clapping)	さんさ時雨	さんさしぐれ	1964
Sanseiu (Acid Rain)	酸性雨	さんせいう	1712

Sanzokuame (A Sudden Rainstorm with Thunder That Comes So Quickly You Have No Time to Tie Even Three Bundles of Rice)	山賊雨	さんぞくあめ	1965
Sanzokuame (Three Bundles Rain / After a Thunderstorm, You Have Time to Wrap Three Bundles of Rice Plants before the Rain Begins Again)	三束雨	さんぞくあめ	1569
Sanzui (Rainy Season—Toyama)		さんずい	1395
Saotate (A Vertical Rain with Strong Wind—Miyazaki)		さおたて	1966
Saran (Stormy Weather during the Rainy Season)	沙乱	さらん	1630
Saruzake (Monkey Liquor—Berries That Monkeys Had Collected in Trees, Fermented and Mixed with Rain and Were Found and Enjoyed by Hunters and Lumberjacks)	猿酒	さるざけ	1967
Saryū (Straw Raincoat)	蓑笠	さりゅう	1285
Sāsā (The Sound of a Rain Shower)	サーサー	さーさー	1690
Satsukiame (The Name of the Fifth Lunar Month / A Long Rain That Falls in the Fifth Lunar Month)	五月雨	さつきあめ	1373
Satsukibare (Clear and Fine Weather despite Being in the Rain Season of the Fifth Lunar Month)	五月晴	さつきばれ	1087
Satsukigawa (May Rain River / A River Swollen by the Continuous Rain of the Month of May)	五月川	さつきがわ	1993
Satsukiyami (A Dark Night in the Rainy Season)	五月闇	さつきやみ	0154

Satsukizora (Sky of the Running Raining Season / Fifth Lunar Month)	皐月空	さつきぞら	1632
Satto (Sudden Sounds of Rain)		さっと	0353
Satukiame (Persistent and Very Wet Rain in May)	五月雨	さつきあめ	1968
Sau (Rain That Falls on the River Shoal)	沙雨	さう	1151
Sausasei ("You Think It Will Rain and Then the Sky Begins to Clear"—Zengo)	乍雨乍晴	さうさせい	1092
Sawake (Thin and Sparse, Faint Rain That Falls Just before a Heavy Downpour—Koza Region, Kanagawa)		さわけ	1089
Sayoshigure (A Weak and Light Rain Shower in the Night)	小夜時雨	さよしぐれ	0168
Sayoshigure (The First Rain That Fell This Year)	小夜時雨	さよしぐれ	0384
Sazanka Chirashi (Rain That Washes the Camellia Flowers Away)	山茶花ちらし	さざんかちらし	1200
Sazanka Shigure (The Rain That Falls on Mountain Flowers / Winter)	山茶花時雨	さざんかしぐれ	0089
Sazankazuyu (Camellia Rain / A Steady Rain in Early Winter)	山茶花梅雨	さざんかづゆ	1578
Sazui (Early Summer Rain)		さずい	1088
Seburi (Rain Deep in the Mountains—Nara)		せぶり	1474
Seifū Ichijinn Kitaru (The Western Winds Begin to Blow, Signaling It Will Rain Soon / September)	西風一陣来	せいふういちじんきたる	1090

Seifū Kuu (Strong Wind, Long Rain)	凄風苦雨	せいふうくう	1635
Seigetsu (Moon Shining in a Rain-Cleansed Sky)	霽月	せいげつ	0683
Seikō Udoku (Cultivating the Fields on Sunny Days and Staying Indoors on Rainy Days to Read Books / Living a Happy Life)	晴耕雨読	せいこううどく	0473
Seikō Uki (A View That Is Beautiful Whether It Is Sunny or Rainy)	晴好雨奇	せいこううき	0030
Seiran (Blue Storm)	青嵐	せいらん	0288
Seiri Usan (Gathered Things That Disperse Quickly, Separated like Clouds and Fall like Rain)	星離雨散	せいりうさん	0047
Seiro (Rain)	清露	せいろ	0972
Seiu (Clean and Pure Rain)	清雨	せいう	1183
Seiu (Quiet and Soft Drizzle)	静雨	せいう	1128
Seiu (Rain That Falls on Leaves)	青雨	せいう	0835
Seiu (Severe Rain / Cold and Depressing Rain)	凄雨	せいう	1636
Seiu (Star Rain / Meteor Shower, Stars Falling like Rain)	星雨	せいう	0646
Seiu (The Rain Stops and the Sun Comes Out)	霽雨	せいう	1969
Seiu (To Pray to God or Deities or the Buddha for Rain)	請雨	せいう	1477
Sekiran'un (Rain Cloud)	積乱雲	せきらんうん	0112
Sekisetsu (Hail and Snow Accumulated on the Ground)	積雪	せきせつ	1320
Sekison (Shrine with a Large, Holy Stone That Brings Rain as Its Principal Object of Worship in Mount Oyama)	石尊	せきそん	1063

Sekiu (Calm Evening Rain)	夕雨	せきう	1475
Sekiu (Long Rain)	積雨	せきう	1552
Semishigure (Cicada Rain / May Rain)	蝉時雨	せみしぐれ	1554
Senchūu (Rain on March 10 That Cleans the Emperor's Kitchen after the Party)	洗厨雨	せんちゅうう	1567
Sendataki (Lighting Bonfires in the Mountains as a Ritual to Pray and Beg the Deities for Rain)	千駄たき	せんだたき	1462
Sengaiu (Rain on March 8, the Rain Cleans the City before the Emperor's Birthday Party Guests Arrive)	洗街雨	せんがいう	1694
Sengetsu (The Moon after the Rain / A Clear and Refreshed Mind)	繊月	せんげつ	1970
Senjō Kōsuitai (Rain Clouds)	線状降水帯	せんじょうこう すいたい	1555
Senpatsuu (Rain on August 16 That Washes the Bon Festival Implements)	洗鉢雨	せんぱつう	1562
Senshyau (Rain That Falls on the Day before Tanabata, July 6 / The Water That Washes the Ox Carriage Used by Hikoboshi to Meet Orihime)	洗車雨	せんしゃう	0228
Senten'u (Autumn Rain)	詹天雨	せんてんう	1202
Setsusetsu (Fine Rain Falling)	屑屑	せつせつ	1794
Setsuu (Fine and Delicate Rain)	屑雨	せつう	1564
Setsuu (Rain Pouring Down on You)	渫雨	せつう	1568
Shadare (Rain Falling Sideways)		しゃだれ	1987
Shajiku o Kudasu (It Rains Heavily)	車軸を下す	しゃじくをくだす	0775

Shajiku o Nagasu (Rain That Makes the Axles Float / Very Heavy Rain)	車軸をながす	しゃじくをながす	1710
Shakyaku (Light and Rain Coming Down Diagonally)	斜脚	しゃきゃく	1899
Shaō no Ame (Rain on Spring Festival That Is Brought by a Local Deity)	社翁の雨	しゃおうのあめ	1565
Shau (Rain That Falls Sideways and Diagonally, Carried by Strong Wind)	斜雨	しゃう	1166
Shibaame (Intermittent Heavy Rain)	繁雨	しばあめ	1566
Shibaame (Rain That Falls Intermittently)	屢雨	しばあめ	0221
Shibakureame (Rain That Falls Heavily)	柴榑雨	しばくれあめ	1681
Shibishibi (Sound of Drizzling Rain—Kyoto)	シビシビ	しびしび	1760
Shiboshiboame (Weak and Faint, Sparse Rain—Niigata)		しぼしぼあめ	1570
Shibukiame (Rain with Strong Wind)	繁吹き雨	しぶきあめ	1679
Shibuku (Rain and Wind Blows Hard)	繁吹く	しぶく	1473
Shibushibuame (Weak and Sparse Rain—Azuma, Gunma)	しぶしぶ雨	しぶしぶあめ	0877
Shidaraden (Loud Thunder and Lightning Caused by Rain)		しだらでん	0878
Shigure (A Late Autumn and Early Winter Rain / To Shed Tears)	時雨	しぐれ	1994
Shigure (Cold Rain, Sleet, and Drizzle in Late Autumn to Early Winter)	時雨	しぐれ	1169

Shigure Gokochi (Feels like Drizzle)	時雨心地	しぐれごこち	0322
Shigure no Aki (Autumn Rain)	時雨の秋	しぐれのあき	1488
Shigure no Ame (Brief, Weak, Sparse, Light Rain from Late Autumn to Early Winter)	時雨の雨	しぐれのあめ	0703
Shigure no Iro (Rain-Colored Grass and Trees)	時雨の色	しぐれのいろ	1773
Shigure no Iro (The Color of the Rain)	時雨の色	しぐれのいろ	1747
Shigure no Matsu (Pine Tree Rain Drizzle / The Middle of the Raining Season)	時雨の松	しぐれのまつ	1971
Shigure no Somuru Yama (Mountain-Dyed Rain)	時雨の染むる山	しぐれのそむるやま	1650
Shigure Wataru (Tears Spill like Drizzling Rain / A Brief and Passing Rain Shower)	時雨渡	しぐれわたる	0325
Shigureakari (A Brief Rain Quickly Passes By and Turns the Sky White and Bright)	時雨明り	しぐれあかり	1153
Shigurebachi (Rain Prayer Bowl Used in Rainmaking Rituals)	時雨鉢	しぐればち	1579
Shiguregasa (Umbrella for Light and Drizzling Rain of Early Winter)	時雨傘	しぐれがさ	1315
Shiguregumo (Winter Drizzle Clouds)	時雨雲	しぐれぐも	0285
Shigureguse (When Rain Increases and Grows Heavier during the Raining Season as if the Rainclouds Were Developing a Habit)	時雨癖	しぐれぐせ	1099

Shigureki (Anniversary of Winter Drizzle / October 12, the Death Day of Bashō, Who Died in the Season of Shigure, and This Word Embodies His Poetry)	時雨忌	しぐれき	1902
Shigureki (The Time of Rain / Winter)	時雨忌	しぐれき	0318
Shiguremichi (Rainy Road)	時雨みち	しぐれみち	1957
Shigureniji (A Rainbow One Sees after a Very Brief Rain—Metaphor for Ephemerality)	時雨虹	しぐれにじ	1922
Shigureru (To Rain On and Off)	時雨れる	しぐれる	1523
Shigurezuki (The Name for October in the Lunar Calendar, the Month of Late Autumn to Early Winter Rains)	時雨月	しぐれづき	1748
Shiguru (Raining Sporadically)	時雨る	しぐる	1500
Shigururu (Drizzling Sleet Is Falling)		しぐるる	0339
Shikainai Kōu (It Is Not Raining above You, but You Can See the Rain in the Distance)	視界内降雨	しかいないこうう	1913
Shikake (Rainfall / Tottori)		しかけ	1415
Shikeame (Long and Enduring Rain That Moistens an Area)		しけあめ	0879
Shikesamu (The Air Is Moist and Cold Due to Autumn Rain)	しけ寒	しけさむ	1927
Shiki no Ame (Rain of the Four Seasons—National Elementary School Anthem, Japan)	四季の雨	しきのあめ	1955
Shikifuru (Raining a Lot)	頻降る	しきふる	1017
Shimaku (Rain Falling for a While—Shiga)		しまく	1508

Shimer (It Will Rain Moderately)	湿り	しめり	1930
Shimeyakana Ame (Soft and Gentle Rain)	しめやかな雨	しめやかなあめ	0098
Shimidaore (When a Cold Winter Morning Turns to Rain during the Day Instead of Becoming Sunny)	凍倒	しみだおれ	1102
Shimo Tsuyokereba Ame to Naru (If the Frost Is Strong, It Will Bring Rain)	霜強ければ雨となる	しもつよければあめとなる	1928
Shimonagashi (Rain on a Morning with Frost—Iwate)	霜流し	しもながし	1101
Shin-u (Great Rain / Very Heavy Rain)	深雨	しんう	0408
Shin'u (Deep Rain / Heavy Rain)	深雨	しんう	1456
Shin'u (New Rain / Rain That Falls at the Time of Fresh, New Green)	新雨	しんう	1201
Shinai (Rain—Aizu Region, Fukushima)		しない	1199
Shindō raiden (Lightning and Thunder Caused by Wind and Rain, Very Noisy)	震動雷電	しんどうらいでん	0893
Shindou (Thunderstorm)	瞋怒雨	しんどう	1100
Shinogu (To Stave Off Rain)	凌	しのぐ	1281
Shinotsukuame (Intense Rain That Falls Heavily, Is Very Fine and Strong like the Bamboo Grove at Shinotake)	篠突く雨	しのつくあめ	0298
Shinryū (Morning Raindrops)	晨霤	しんりゅう	1062
Shinsen'en (Sacred Spring Garden Located near Nijo Castle in Kyoto Where the Law of the Rain Sutra Was Practiced after a Miraculous Rain Fell after Monk Kūkai's Rain Prayers)	神泉苑	しんせんえん	1863

Shinzen'en (Temple Where the Priest Eun Performed the First Recorded Rainmaking Ritual in 854)	神泉苑	しんぜんえん	1594
Shion no Ame (Rain That Falls during a Warm Spring; after Three Cold Days, Four Warm Days Will Follow)	四温の雨	しおんのあめ	1711
Shioshio (Wet with Tears or Wet with Rain)		しおしお	1915
Shiposhipo Ame (Moist Rainfall—Niigata)	しぽしぽ雨	しぽしぽあめ	1145
Shippori (Rain That Falls Quietly)		しっぽり	1509
Shippū (Humid Southern Wind Bringing Rain and Wet Air from the Ocean)	湿風	しっぷう	0429
Shippū Yokuu (Comb Your Hair with the Wind and Wash Your Body in the Rain / Exposed to the Elements / Struggling)	櫛風浴雨	しっぷうよくう	0233
Shirasame (White Rain, Another Name for an Evening Shower)	白雨	しらさめ	1774
Shirobae (White Glow Rain / During the Rainy Season, the Sky Seems to Become Bright and Clear with Just a Little Fine Rain)	白映え	しろばえ	1671
Shiroku Hikui Kumo ga Mine ni Kakaru Toki wa Sūjitsu no Uchi ni Ame ni Naru (When Low White Clouds Hit the Peak, It Will Rain after a Few Days' Time)	白く低い雲が峰にかかるときは、数日のうちに雨になる	しろくひくいくもがみねにかかるときは、すうじつのうちにあめになる	0425
Shishiraburu (Rain Spraying Finely—Oita)		ししらぶる	1745
Shitoburu (A Very Brief Rain—Yamagata)		しとぶる	1670

Shitoshito (The Sound of Rain Resonating in a Quiet Place)		しとしと	0207
Shitoshito Ame (Spotty Rain)	しとしと雨	しとしとあめ	0031
Shitsuu (Comb Rain / To Arrange Your Hair in the Rain / To Work Hard Outside)	櫛雨	しつう	0699
Shitsuu (Rain That Falls Heavily)	疾雨	しつう	1144
Shiu (It Is Raining Suddenly / Fast Rain)	駛雨	しう	0583
Shiu (Praying Rain / Praying for the Rain to Stop)	止雨	しう	0064
Shiu (Thin Rain like a Thread)	糸雨	しう	0443
Shiyashiya (Sound of Rain)		しやしや	1759
Shizuka Gozen (Lady Shizuka Gozen [1165–1211] Danced during a Rain Prayer Ceremony at Shinsen-en Garden and Immediately Brought Dark Clouds and Rain after 100 Monks Praying for Rain Had Been Unsuccessful)	静御前	しずかごぜん	1585
Shizuku (Droplet)	雫	しずく	0281
Shobofuru (Rain Drizzling Down)	しょぼ降る	しょぼふる	1511
Shobokeame (Faint Rain—Shimane)		しょぼけあめ	0639
Shobonureru (To Get Drenched by Rain)	しょぼ濡れる	しょぼぬれる	1914
Shoboshobo (Drizzling, Very Faint Rain)		しょぼしょぼ	0395
Shoboshoboame (Steadily Drizzling Rain)	しょぼしょぼ雨	しょぼしょぼあめ	1675
Shobotsuku (To Rain Slowly)		しょぼつく	1418

Shōbunagashi (A Long Spell of Rain during the Rain Season—Miyagi)		しょーぶながし	1887
Shōen Dan'u (Smoke Rain / The Battleground Is Heated and Bullets Fall like Rain)	硝煙弾雨	しょうえんだんう	0059
Shōgetsuu (Rain Falling after the Full Moon)	翔月雨	しょうげつう	0463
Shōshō (Wistful and Sad Feeling of Rain)	蕭々	しょうしょう	1146
Shōu (A Depressing Rain of Autumn)	蕭雨	しょうう	1292
Shōu (A Little Bit of Light Rain)	小雨	しょうう	1290
Shōu (Light and Gentle Rain)	霎雨	しょうう	1284
Shōu (Night Rain)	宵雨	しょうう	1693
Shōu (Praying for Rain)	請雨	しょうう	0344
Shōu (Rain on the Pine Trees)	松雨	しょうう	1340
Shōu (Rain with Vapor)	瘴雨	しょうう	0584
Shōu (The Rain That Falls during the Heat and Humidity of Summer)	暑雨	しょう	1155
Shōu kekkō (Canceled by Heavy Rain)	小雨決行	しょううけっこう	0725
Shōuhō (Rain-Praying Ritual for Times of Drought / Esoteric Buddhism)	請雨法	しょううほう	1756
Shōukyō no Hō (The Rain Sutra)	請雨経法	しょううきょうの ほう	1916
Shōukyōhō (Rainmaking Prayer Famously Made by Kobo Daishi [Kukai] at Shinsen-en Garden in 1240)	請雨経法	しょううきょうほう	0373
Shōyō (A Legendary Bird That Dances When It Is About to Rain)	商羊	しょうよう	1874

Shūame (Rain Falling from Clouds)	しゅう雨	しゅうあめ	0111
Shūchūgōu (Strong Downpour in One Area / Localized Shower)	集中豪雨	しゅうちゅうごうう	0354
Shukuu (Rain That Has Continued to Fall since the Night Before)	宿雨	しゅくう	0005
Shun'u (Gentle Spring Rain)	春雨	しゅんう	0208
Shuniji (Main Rainbow / Primary Rainbow)	主虹	しゅにじ	0801
Shunpū Kōu (Rain with a Warm Spring Breeze That Nourishes and Enriches the Crops)	春風膏雨	しゅんぷうこうう	1917
Shunran (Spring Storm)	春嵐	しゅんらん	0374
Shunrin (A Fine Rain That Continues to Fall during Spring)	春霖	しゅんりん	0146
Shunrin (Misty Spring Rain)	春霖	しゅんりん	0253
Shūren Kiryūsei Kōu (Rain Caused by Horizontal Wind and Air)	収斂気流性降雨	しゅうれんきりゅうせいこうう	1735
Shūrin (Autumn Rain)	秋霖	しゅうりん	0475
Shūrin (Long, Enduring Rain)	修霖	しゅうりん	1920
Shūrinzensen (Autumnal Rain Front)	秋霖前線	しゅうりんぜんせん	1293
Shuroshuro (Non-stopping and Gloomy, Weak Rain—Tamana Region, Saitama)		しゅろしゅろ	0638
Shussui (Flooding from the Rain)	出水	しゅっすい	1641
Shutsueki (When the Rainy Season Ends)	出液	しゅつえき	0637
Shutubai (Plums Leaving / The End of the Rainy Season)	出梅	しゅつばい	1918
Shūu (A Sudden Shower / Cloudburst)	驟雨	しゅうう	0025

Shūu (A Very Long Rain for Three Days in a Row)	衆雨	しゅうう	1147
Shūu (Cold Rain in Autumn)	秋雨	しゅうう	0431
Shūu (Wistful Rain That Saddens People, Especially during the Seasonal Change between Autumn and Winter)	愁雨	しゅうう	1291
Shūu ryūshun (Light Spring Rain)	小雨留春	しょううりゅうしゅん	1535
Sobae (To Rain While the Sun Is Shining / The Marriage of a Fox / Summer)	日照雨	そばえ	0975
Sōbai (Rain Shower—Hyogo)		そーばい	1581
Sōbaiu (A Heavy Summer Rain That Falls at the End of Rain Season)	送梅雨	そうばいう	1296
Sobofuru (Steady Rain)	そぼ降る	そぼふる	0176
Sobofuruame (Smooth and Calm Rain)	そぼ降る雨	そぼふるあめ	1154
Sobosobo (The Way the Rain Falls Softly)		そぼそぼ	0014
Sobotsu (Rain That Falls Elegantly and Steadily)	濡	そぼつ	0794
Sobotsu (Wet and Soaked by Rain)	濡つ	そぼつ	0034
Sōburu (Silent and Gentle Rain)		そおぶる	0287
Sode no Ame (Rain That Wets the Sleeves / Tears of Sorrow Falling on the Sleeves)	袖の雨	そでのあめ	1586
Sodegasa Ame (Rainfall That Is Light and Not So Wet, You Can Use Your Sleeve as an Umbrella)	袖笠雨	そでがさあめ	0082
Sodeshigure (Sleeve Rain / Tear on a Kimono Sleeve)	袖時雨	そでしぐれ	1704

Sōfuru (To Rain Gracefully)	そお降る	そおふる	1941
Soga no Ame (Soga Brothers Avenged on a Rainy Day, May 28)	曽我の雨	そがのあめ	0294
Sohobaru (Gentle Rain)	添降	そほぶる	0273
Sohobaru (Graceful and Slow Soft Rain—Ise Monogatari)	添雨	そほぶる	1187
Sohobaru (Rain That Falls Quietly and Gently)		そほぶる	0343
Sohobaru (Soft, Slow Rain)	壮雨	そほぶる	1607
Sohofuru ("It Rains Gracefully" —Song from the Country of Koshi-nomichi No Naka, Man'yōshū [Collection of Ten Thousand Leaves], Volume 16, Song 3883)	曾保零 (万葉集 第十六巻 三千八百八十三 番歌) 越中(こし のみちのなか) の国の歌。	そほふる (まんよう しゅう だい16かん 3883ばんか) えっちゅう (こしの みちのなか) のくに のうた。	0836
Sohofuru (It Rains Slowly)	そほ降る	そほふる	1541
Sohotsu (Rain or Tears That Fall Gracefully)		そほつ	1257
Sohotzu (To Become Wet with Rain or Tears)		そほづ	0434
Sokuu (Raining Enough)	足雨	そくう	1294
Sora Ni Mitsurouka (Three Corridors in the Sky / Is It Rain, Shine, or Cloud)	空に三つ廊下	そらにみつろうか	1298
Sora no Shigure (Rain from the Sky—Metaphor for Tears)	空の時雨	そらのしぐれ	1170
Sorashiranuame (Tears Are the Rain That Does Not Fall from the Sky / Tears Are Rain That the Sky Does Not Know)	空知らぬ雨	そらしらぬあめ	1295
Sōtottesōtotte (Make It Rain, Make It Rain!—Chanted during Rain-Invoking Prayer Rituals Begging the Deities for Rain)		そーとって、そー とって	1929

Sōtsu (Raining Gracefully)		そおつ	1972
Sōu (A Sudden Heavy Downpour of Rain)	早雨	そうう	1297
Sōu (It Rained Heavily, Then Stopped and Started Again)	叢雨	そうう	1921
Sōu (Rain Falling on a Window)	窓雨	そうう	0611
Sōu (Scattered, Drizzling Rain)	疎雨	そう	0177
Sōu (Sparse Rain)	疏雨	そう	0290
Sōu (Sparse, Scattered Rain)	疎雨	そう	1352
Sowoburu (Elegant Rain)	ソヲブル	そをぶる	0388
Sowofuru (Elegant and Quiet Rain)	ソヲフル	そをふる	0131
Sozoroame (Lightly Falling Rain That Never Stops)	漫ろ雨	そぞろあめ	1148
Sōzu (Silent and Steady Rain)		そおづ	1702
Ssanshūami (Autumn Rain— Ikema Region, Okinawa)		っさんしゅうあみ	1300
Suijin (Deity of Water Prayed to during Rain-Praying Festival at Kifune Shrine)	水神	すいじん	1647
Suiu (Rain That Falls on Leaves)	翠雨	すいう	1584
Sukunabikona No Mikoto (The Deity of Rain and Medicine)	少彦名命	すくなびこなの みこと	0412
Sūmanbōsū (Rainy Season—Okinawa)		すーまんぼーすー	1868
Sunekosuri (A Phantom or Specter That Appears during the Rain and Looks like a Puppy)		すねこすり	1359
Suneoriamagoi (Rain-Praying Festival with a Two-Ton Dragon Carried from a Shrine into a Lake, Dismantled in the Water to Anger Deities to Cause Rain—Saitama)	脚折雨乞	すねおりあまごい	1557

Susanoo (Deity Named "Raging Storm" Who Presides over Storms and Rainstorms)	素戔男尊	すさのお	1494
Susu ga Ochiru to Ame (When the Soot Falls, Rain)	煤が落ちると雨	すすがおちるとあめ	1149
Susukibaiu (Rainy Season on the Pampas Grass)	すすき梅雨	すすきばいう	1923
Syūu (Cold Rain in Autumn)	驟雨	しゅうう	0397
Tabibito to Wa ga Na Yobaren Hatsu Shigure ("I Am Called the Traveler, First Winter Drizzle" —Matsuo Bashō)	旅人と我が 名呼ばれん 初時雨 (芭蕉)	たびびと とわがなよ ばれん はつしぐれ (ばしょう)	0724
Tachiame (Drizzle in Tochigi)	たち雨	たちあめ	1709
Tachisame (A Sudden Rain with Droplets That Look as if They Are Standing in Vertical Lines)	立雨	たちさめ	1344
Tadare (Long Rain of Autumn)		ただれ	0843
Tade no Ame (Rain Falling on Buckwheat)	蓼の雨	たでのあめ	0391
Tagarāme (Rain Coming after a Summer Drought)		たがらーめ	1471
Taifū (Great Rain / Very Heavy Rain)	台風	たいふう	0099
Taifūikka (Clear Weather after the Typhoon Passes)	台風一過	たいふういっか	0258
Taikan Jiu (Mercy-from-Drought Rain / To Wait a Long Time or Eagerly Wait in Times of Difficulty as When One Waits for Rain during a Drought)	大旱慈雨	たいかんじう	0950
Tairyūseikōu (Rain Caused by Temperature Change in the Sky)	対流性降雨	たいりゅうせい こうう	0689
Taisōu (Rain during the Season)	太宗雨	たいそうう	1150

Taiu (Heavy Rain)	大雨	たいう	1717
Taiu (Rain That Falls in Large Quantities)	大雨	たいう	1998
Taiu (To Face the Rain)	対雨	たいう	1999
Taiu (When the Atmosphere Feels a Bit Rainy)	帯雨	たいう	0949
Taiu tokidoki Furu (Great Rains Sometimes Fall, Season Thirty-six in Seventy-two Kō Ancient Calendar)	大雨時行	たいうときどきふる	0455
Taiurin (Long Rain for Three Days)	大雨霖	たいうりん	0726
Taiyō no Iro ga Usuku Mieru Toki wa Ame no Shirushi (When the Color of the Sun Appears Lighter, Rain)	太陽の色が薄く見えるときは、雨のしるし	たいようのいろがうすくみえるときは、あめのしるし	0800
Takanuami (Calm Autumn Rain)		たかぬあみ	1542
Takanushībai (Hawk Rain / Fog-like Rain in the Early Autumn Season, When Hawks Are Migrating South)		たかぬしーばい	0903
Takaokami-No-Kami (Deity of Water and Rainfall)	高龗神	たかおかみのかみ	1533
Takaraame (Rain That Is a Treasure / Joyful Rain)	宝雨	たからあめ	0110
Takawatari (Long Enduring Rain at the End of September—Miyazaki)		たかわたり	0947
Take no Ko Nagashi (Rain during Bamboo Season / Rain That Washes Bamboo Shoots)	筍流し	たけのこながし	0841
Take no Nobori (Climbing the Mountain Peak as a Rain-Praying Ritual Event—Nagano Besso Jinja)	岳の幟	たけののぼり	1864
Takenokotsuyu (Rain That Falls on Bamboo Shoots)	たけのこ梅雨	たけのこつゆ	0534

Takiotoshi (Rain like a Waterfall Dropping Water / Heavy Rain)	滝落し	たきおとし	0360
Takitsuhiko (Deity of Pouring Rain and Waterfalls)	多伎都比古	たきつひこ	1865
Takkora Takkora (Strong Rain—Mino and Masuta Regions, Shimane)		たっこらたっこら	0842
Takushiu (Heavy Rain around July That Breaks the Branches off Trees)	濯枝雨	たくしう	1976
Takuu (Blessed Rain That Quenches All Things in the Universe)	沢雨	たくう	1720
Tamadare (Raindrops)	玉垂れ	たまだれ	1470
Tamamizu (A Raindrop / Rainwater Dripping from the Eaves)	玉水	たまみず	1468
Tameoke (Rainwater-Collecting Bucket)	ため桶	ためおけ	1472
Tamoto no Shigure (Rain on Sleeves / A Kimono Sleeve Wet from Wiping away Tears)	袖の時雨	たもとのしぐれ	0462
Tan'u (Morning Rain)	旦雨	たんう	1469
Tanabata Nagashi (Rain Flowing on July 7)	七夕流し	たなばたながし	0691
Tanabataame (Rain That Falls during the Tanabata Star Festival Night of July 7)	七夕雨	たなばたあめ	0129
Tangetsu (The Moon after It Rains)	潭月	たんげつ	1974
Tanomu Ki no Shita ni Ame Moru (I Asked a Tree for Shelter from the Rain but Drops Still Fell on Me)	頼む木の下に雨漏る	たのむきのしたにあめもる	0096
Tanrokujiu (Loud, Heavy Rain)	湛轆耳雨	たんろくじう	1977
Tantekiame (A Sudden Rain—Awa Region, Chiba)	端的雨	たんてきあめ	1467

Tatekōzui (Torrential Rain—Yamanashi)	たて洪水	たてこうずい	1008
Tatekōzui (Vertical Flood / Extremely Heavy Rain)	縦洪水	たてこうずい	0710
Tateshinadachi (Evening Rain Coming from the Direction of Tateshina Mountain)	蓼科立	たてしなだち	1347
Tatsumi (The Direction between the Dragon and the Snake / Strong Southeast Winds)		たつみ	0690
Tau (Heavy Rain)	多雨	たう	0166
Taueame (Rain Season / Shimane)	田植雨	たうえあめ	1453
Tauesazui (Rain Season—Mino and Masuta Regions, Shimane)	田植さずい	たうえさずい	1696
Te o Hirugaeseba Kumo to Nari Te o Kutsugaeseba Ame to Naru (Turn Your Hand, It Becomes a Cloud; Cover Your Hand, It Becomes Rain)	手を翻せば雲となり、手を覆せば雨となる	てをひるがえせばくもとなり、てをくつがえせばあめとなる	0155
Teikiatsuseikō (Sleet and Rain, Frontal Rain)	低気圧性降雨	ていきあつせいこうう	1345
Teīrāmi (A Sudden Sprinkle of Rain—Okinawa)		てぃーらあーみ	1416
Teishunakase no Ame (Rain That Makes the Masters Cry / When It Rains Only during the Day and Stops at Night)	亭主泣かせの雨	ていしゅなかせのあめ	0787
Ten no Mizu (Water from the Sky / Rain)	天の水	てんのみず	0390
Ten'u (When It Rains on Time)	天雨	てんう	0711
Tenchōukō (Rain with the Sun Directly over Your Head)	天頂降雨	てんちょうこうう	1697

Tengai ga Bareru (Heaven's Canopy Is Exposed / It Will Rain)	天蓋がばれる	てんがいがばれる	1698
Tenkiame (It's Sunny but Raining / Fox Rain)	天気雨	てんきあめ	0303
Tenkyū (There Are No Clouds in the Sky but a Fine Rain Is Falling, a Wonderful and Delicate Rain as if Heaven Is Crying)	天泣	てんきゅう	0512
Tennon (Rain of Grace from Heaven)	天恩	てんのん	1589
Tensoku (Rain That Has Poured Enough)	霑足	てんそく	1700
Tensui (Water from the Sky / Water of the Heavens / Rain)	天水	てんすい	0247
Tensui nōgyō (Water from Heaven / Rainwater Agriculture)	天水農業	てんすいのうぎょう	1664
Tensuiba (Water from the Sky Field / Rice Farming That Relies on Rainwater)	天水場	てんすいば	1389
Tensuiden (Heaven's Water Rice Field / Rainwater Rice Cultivation)	天水田	てんすいでん	1665
Tensuioke (Looking into the Rainwater Tank, Collecting the Stormwater from Heaven)	天水桶	てんすいおけ	0045
Tenteki (Raindrop)	点滴	てんてき	0308
Teriame (Raining While the Sun Is Shining)	照り雨	てりあめ	1624
Terifuri (Fox Bride Rain / Raining While Sunny)	照り降り	てりふり	1386
Terifuriame (Rain That Is Changeable and Comes Down When You Think It Is Clear but Stops When You Think It Should Rain)	照降雨	てりふりあめ	1981

Terifurinashi (No Uncertain Rain / With Absoluteness Regardless of Rain)	照降無	てりふりなし	1867
Teritsuyu (Clear Skies and Very Little Rain during Tsuyu, the Rainy Season)	照り梅雨	てりつゆ	0044
Teruteru Bōzu (Sunny Monk, a Magical Talisman Made with White Cloth That Farmers Hung in Their Windows to Stop the Rain)	照る照る坊主	てるてるぼうず	1978
Tirabui (Fox Bride Rain / Rain in the Sunshine)		てぃらぶい	1979
Tobishyari (Splashes of Rainwater)		とびしゃり	1727
Tōjiame (Rain on Winter Equinox)	冬至雨	とうじあめ	1544
Tōka no Ame (Rain in August to September When the Soy Flowers Are in Bloom)	豆花の雨	とうかのあめ	1708
Tōka no Ame (Rain on Peach Blossoms)	桃花の雨	とうかのあめ	1716
Tōkau (Autumn Rain)	豆花雨	とうかう	0020
Toki no Ame (Rain around Autumn and Winter)	時の雨	ときのあめ	1707
Tokidoki Niwaka Ame (Sometimes Light Snow and Rain Showers)	時々にわか雨	ときどきにわかあめ	1045
Tokidoki Niwaka Yuki (Sometimes Snow or Sometimes Light Snow or Rain)	時々にわか雪	ときどきにわかゆき	0156
Tokidokiame (Temporary and Brief Rain That Ends Quickly)	時々雨	ときどきあめ	1798
Tokishiru Ame (Rain to Know the Time / Gentle Rain That Stops as Soon as It Starts, an Umbrella Is Not Necessary)	時知る雨	ときしるあめ	0461

Tokoroburi (Evening Shower in a Certain Area)	所降り	ところぶり	1866
Tomisagari (Rain of New Year's That Brings Fortune and Wealth)	富下り	とみさがり	1387
Tomishōgatsu (Rain on the First Day of the Year That Brings Wealth)	富正月	とみしょうがつ	1390
Tora ga Namida (Tiger Rain on May 28)	虎が涙	とらがなみだ	1608
Toraga Ame (The Tiger Is in Tears / Rain on May 28 of the Lunar Calendar)	虎が雨	とらがあめ	0254
Tōriame (Brief Rain Flurry That Passes Quickly By)	通り雨	とおりあめ	0060
Tōu (Freezing Cold Rain and Frozen Raindrops / Winter)	凍雨	とうう	0381
Tōu (To Pray for Rain)	禱雨	とうう	1648
Tōu (Winter Rain)	冬雨	とうう	1448
Toūkagushinu Yusami (It Has Been Raining since Ten Days Ago)		とぅかぐしぬゆさみ	1527
Tsubame ga Hikuku Tobu to Ame (When the Swallows Fly Low, Rain)	ツバメが低く飛ぶと雨	つばめがひくくとぶとあめ	0483
Tsubo (The Direction in Which the Evening Raintorm Always Arrives—Kumamoto)		つぼ	1980
Tsuchi no Shō Uruoi Okoru (Rain Moistens the Soil, Season Four in Seventy-two Kō Ancient Calendar)	土脉潤起	つちのしょううるおいおこる	0400
Tsuchifuru (Dirt and Earth Rain / Dust Storm)	土降る	つちふる	1749
Tsuiri (Around June 11, the Start of the Rainy Season)	入梅	ついり	0240

Tsuiri (Rain That Causes the Chestnut Flowers to Fall / To Enter the Rainy Season)	堕栗花	ついり	0482
Tsuiri (Rainy Season Begins)	梅雨入り	ついり	1606
Tsuiriame (Chestnut Flower Rain / The Rainy Season)	栗花落雨	ついりあめ	1654
Tsuiribare (Temporary Sunny Day during the Rainy Season)	梅雨入晴	ついりばれ	0722
Tsuiribare (The Rainy Season Is Over and the Weather Will Be Fine)	入梅晴	ついりばれ	0077
Tsuiu (Falling Rain)	墜雨	ついう	1644
Tsuki ga Kasa o Kaburu to Ame (A Halo around the Moon, Rain)	月が暈をかぶると雨	つきがかさをかぶるとあめ	0517
Tsuki ni Amagasa Higasa Nashi (The Moon Has No Rain Hat nor Sun Hat)	月に雨笠日笠なし	つきにあまがさひがさなし	1409
Tsuki no Ame (A Moon Hidden by Rain)	月の雨	つきのあめ	1341
Tsukigasa (Halo of Ice around the Moon)	月暈	つきがさ	1447
Tsukimizutsuki (May, the Month When Rain Makes the Moon Invisible)	月不見月	つきみずつき	0336
Tsukimizutsuki (The Month When There Is No Moon / The Moon Is Obscured by the Rainy Season / May in the Ancient Lunar Calendar)	月見ず月	つきみずつき	0081
Tsukishigure (It Rains When the Moon Is Rising)	月時雨	つきしぐれ	0450
Tsukishigure (Rain in the Moonlight)	月時雨	つきしぐれ	0222

Tsukuriame (Made Rain / In Olden Times, an Occupation Was to Create False Rain by Climbing on Roofs and Pouring Water from Above)	作り雨	つくりあめ	0798
Tsumetai Ame (Cold Rain)	冷たい雨	つめたいあめ	0083
Tsurugashima Ryūjin-sai (August Rain-Praying Festival Honoring the Rain Dragon God Ryūjin)	鶴ヶ島 龍神祭	つるがしまりゅうじんさい	1795
Tsutsunuke (Raining like a Leak)	筒抜け	つつぬけ	1796
Tsuya (Rainy Night)	雨夜	つや	0423
Tsuyu (Rainy Season / Rain Continues June to Mid-July)	黴雨	つゆ	0367
Tsuyu no Chō (Butterflies That Flutter between Rainfall)	梅雨の蝶	つゆのちょう	1548
Tsuyu no Hare (Sunny during Raining Season)	梅雨の晴	つゆのはれ	1750
Tsuyu no Hashiri (A Rainy Spell Just before the Rainy Season)	梅雨の走り	つゆのはしり	0510
Tsuyu no Hoshi (Stars That Are Visible between the Clouds or on Clear Nights during the Rainy Season)	梅雨の星	つゆのほし	1339
Tsuyu no Kehai (When It Feels like the Raining Season Is Arriving Soon)	梅雨の気配	つゆのけはい	1103
Tsuyu no Rai (Thunder during the Rain Season)	梅雨の雷	つゆのらい	1713
Tsuyu no Sora (Sky of the Rainy Season)	梅雨の空	つゆのそら	0518
Tsuyu no Tsuki (The Moon during Raining Season / Moonlight on a Rainy Night)	梅雨の月	つゆのつき	1336

Tsuyu no Yama (Mountains of the Rainy Season, Green and Covered with Rain)	梅雨の山	つゆのやま	1658
Tsuyuagari (The Rain Season Has Lifted and Ends)	梅雨上	つゆあがり	1723
Tsuyuake (End of the Rain Season, Middle of July of Lunar Calendar)	梅雨明け	つゆあけ	1714
Tsuyuana (Rain Season Hole / Holes That Appear from Rainwater on the Road)	梅雨穴	つゆあな	1490
Tsuyuaoi (Plum Rain Flower / The Rainy Season Flower That Blooms at the Beginning of Rainy Season and Finishes Blooming at the End of Rainy Season / Hollyhock)	梅雨葵	つゆあおい	1849
Tsuyubare (Sunny Rainy Season)	梅雨晴れ	つゆばれ	1337
Tsuyudemizu (When the Rivers Flood during Raining Season)	梅雨出水	つゆでみず	1715
Tsuyudoki (Wet Season / Rainy Season)	梅雨時	つゆどき	0309
Tsuyugomori (To Stay Inside during Rainy Season)	梅雨籠	つゆごもり	1452
Tsuyugōu (Heavy Rain in the Latter Half of the Raining Season)	梅雨豪雨	つゆごうう	1721
Tsuyugumo (Plum Rain Cloud / Rainclouds of the Raining Season)	梅雨雲	つゆぐも	1478
Tsuyugumori (Cloudy Sky in the Rainy Season)	梅雨曇	つゆぐもり	0282
Tsuyuharema (A Spell of Sunny Weather during the Rain Season)	梅雨晴れ間	つゆはれま	1171
Tsuyuharema (The Rain Season Will End Soon and It Will Be Sunny)	梅雨晴間	つゆはれま	1613

Tsuyuhareru (Sunny during Rainy Season)	梅雨晴る	つゆはれる	1726
Tsuyuiri (Entering the Rainy Season / The Beginning of the Rainy Season)	入梅	つゆいり	0406
Tsuyuiri (Rain at the Time of Chestnut Blooming Season)	梅雨入り	つゆいり	1755
Tsuyujimeri (An Area Becomes Very Damp during Rainy Season)	梅雨湿り	つゆじめり	1450
Tsuyukaminari (Thunder during the Rainy Season / Midsummer)	梅雨雷	つゆかみなり	0327
Tsuyukizasu (The Rain Season Nears)	梅雨兆す	つゆきざす	1728
Tsuyumeku (The Rainy Season Begins)	梅雨めく	つゆめく	0120
Tsuyunagashi (The Rain Season Is Dragging On)	梅雨流し	つゆながし	1725
Tsuyusamu (Cold and Rainy during the Rain Season)	梅雨寒	つゆさむ	1724
Tsuyushitodo (Continuous Heavy Rain That Drenches People and All Things during the Rain Season)	梅雨しとど	つゆしとど	0203
Tsuyuyami (Rainy Season Darkness Due to the Moon Being Covered by Rain and Clouds)	梅雨闇	つゆやみ	1449
Tsuyuzora (Sky of the Rainy Season)	梅雨空	つゆぞら	0300
Tsuyuzukiyō (Plum Rain Moon Night / A Moon Hiding in Rain and Clouds during the Rainy Season)	梅雨月夜	つゆづきよ	1342
U no Ame (Rain at the Hour of the Rabbit, around 6 AM—Takeno Region, Kyoto)	卯の雨	うのあめ	0967

Ūamī (Heavy Downpour—Okinawa)		うーあみー	1771
Uan (Rain Darkness)	雨暗	うあん	0513
Uan (The Dark Sky of Rain)	雨闇	うあん	1990
Uango (When a Monk Stays Inside the Monastery for Training during the Raining Season to Avoid Unnecessary Killing by Walking on All the Plants and Creatures That Thrive during This Time)	雨安居	うあんご	0888
Ubaodoshi (Strong Wind and Rain Passing through the Plains, Frightening Old Women—Nagasaki)	姥おどし	うばおどし	1161
Ubume (A Female Ghost Who Carries a Baby and Comes Out in the Rain and Tries to Make People Who Pass By Hold Her Baby)	姑獲鳥	うぶめ	1525
Uchiami (Rain Falling toward the Inside of Buildings)	ウチアミ	うちあみ	1642
Uchiku Fūshō Minazen o Toku (Rain, Bamboo, Wind, Pine, All in Zen)	雨竹風松皆説禅	うちくふうしょう みなぜんをとく	2000
Uchisu (Rain That Falls Quickly / Brief Rain)	打ち為	うちす	0190
Uchō (Rainy Morning)	雨朝	うちょう	1543
Uchū (In the Pouring Rain / A Fierce Battle in the Rain)	雨中	うちゅう	1156
Uchū (Non-Stop Pouring Rain)	雨注	うちゅう	1992
Uchū Yurai no Ame (Rain That Comes from Space)	宇宙由来の雨	うちゅうゆらい のあめ	1526
Uda (For the Rain to Hit Something)	雨打	うだ	0023

Udou (Rain Road / Slippery Conditions on the Road Due to Rain)	雨淖	うどう	1610
Uen (Rain Swift Bird / Swallows)	雨燕	うえん	1046
Ugetsu (Rainy Moon / Being Unable to See the Harvest Moon as It Is Obscured by Rain)	雨月	うげつ	0509
Ugetsu Monogatari (*Rainy Moon Tales, a Book of Mysterious Beings Appearing in the Rain and on Mornings with a Lingering Moon* —Akinari Ueda, written in 1776)	雨月物語	うげつものがたり	1338
Ugetsugaku (Rain That Falls on the First Day of the Month)	雨月額	うげつがく	0829
Ugi (Deity of Rain)	雨祇	うぎ	0828
Ugi (The Palace Rituals That Are Omitted When It Rains)	雨儀	うぎ	1157
Ugo (After the Rain)	雨後	うご	0451
Ugo no Hi (Sunshine after a Downpour)	雨後の日	うごのひ	1848
Ugo No Takenoko (Bamboo Shoots after Rain—Parable for Things That Appear One after Another)	雨後の竹の子	うごのたけのこ	0361
Ugo no Takenoko (Bamboo Shoots after the Rain—Proverb for Things Happening One after Another)	雨後の筍	うごのたけのこ	1860
Ugo Shunjyun (Bamboo Shoots after the Rain—Metaphor for Things That Happen Over and Again)	雨後春筍	うごしゅんじゅん	0981
Uha (Frequent Rain)	雨頗	うは	0234
Uhi (The Rain Stops Falling)	雨罷	うひ	1094

Uhi (Things Falling Violently in the Rain)	雨飛	うひ	0191
Uhō Dōji (Divine Rainmaking Child, a Representation of Amaterasu Omikami and Dainichi Nyorai)	雨宝童子	うほうどうじ	0799
Uhyou (Rain Ice)	雨氷	うひょう	0413
Ui (Looks like It Is Going to Rain / Rainy Air)	雨意	うい	1991
Uiki (Rainy Area)	雨域	ういき	0417
Uin (It Rains Too Much)	雨淫	ういん	1791
Uin (Rain Shadow)	雨蔭	ういん	1805
Uiujyō (Rainy Clouds)	雨意雲情	ういうじょう	1792
Uiujyō (Rain and Clouds Affection / The Love and Emotion between a Couple)	雨意雲情	ういうじょう	1512
Ujō (The Feeling of Rain)	雨情	うじょう	0011
Uju (Rain That Enriches the Crops and Plants)	雨澍	うじゅ	1788
Ujun (To Be Enriched by Rain)	雨潤	うじゅん	1789
Uka (Downpour)	雨下	うか	0299
Uka (Flowers That Blossom in the Rain)	雨花	うか	1790
Uka (For the Rain to Pass By)	雨過	うか	0953
Uka (Rain Flower / When Bodhisattvas Gain Great Insight, Heavenly Flowers Fall from the Sky)	雨華	うか	0310
Uka Tensei (The Rain Will Stop, the Skies Will Clear / A Situation or Condition That Was Bad Will Turn Out to Be Good)	雨過天晴	うかてんせい	0079

Ukau (It Rains)	雨降	うかう	1984
Ukei (A Place Where Rainwater Collects in a Gravel Ground)	雨渓	うけい	1787
Ukei (Scenery in the Rain)	雨景	うけい	0954
Uketsu (Blood Rain)	雨血	うけつ	1786
Uketsu (Rain Stops Falling)	雨歇	うけつ	1847
Uki (A Rainy Season for about One Month)	雨期	うき	0952
Uki (Praying for Rain)	雨祈	うき	0887
Uki (Rain Pattern)	雨気	うき	0578
Uki (Rainy Season)	雨季	うき	1520
Uki (Signs of Rain)	雨気	うき	0577
Uki Seikō (A Beautiful Place to Visit in Rain or Sunshine, It Is a Good Thing That the Sun Shines as Much as It Does and a Good Thing It Rains as Much as It Does)	雨奇晴好	うきせいこう	0966
Ukifu (Sky Is Crying / Rain That Appears Where There Are No Clouds and the Heavens Appear to Be Weeping)	雨泣	うきふ	0187
Ukiu (It Has Been Raining for a Long Time)	雨久	うきう	0301
Ukon (Dark Surroundings Due to Rain / The Rain Darkens the Area)	雨昏	うこん	0960
Ukon (Traces of Rain on the Earth, Soil, and Stones)	雨痕	うこん	1638
Ukou (A Mythical Beast That Is Said to Fall with the Thunder and the Rain)	雨工	うこう	0965
Ukou (Rain That Carries the Aroma of Flowers and Earth)	雨香	うこう	1640

Ukou (Rain That Falls on Newly Sprouting Plants)	雨甲	うこう	1599
Uku Wa (Flowers in the Rain / Flowers That Fall in the Rain)	雨花	うくわ	1988
Ukuwa (Rain That Passes Over)	雨過	うくわ	1989
Ukyaku (Streaks of Pouring Rain)	雨脚	うきゃく	0985
Ukyo (The Rain Passing)	雨去	うきょ	1601
Ukyoku (Rain Pole, the World's Rainiest Place)	雨極	うきょく	1639
Ukyū (An Extended Period of Rain)	雨久	うきゅう	1743
Ume no Ame (Rain on the Plums)	梅の雨	うめのあめ	1540
Ume no Tsubuyaki (The Sound of Rain and the Rainy Season)	梅のつぶやき	うめのつぶやき	1492
Umeshigure (Rain When the Plums Are in Season)	梅時雨	うめしぐれ	1785
Umewaka no Namidaame (Tears for Umewaka / A Rain That Falls on March 15 in Sorrow for the Death of a Tragic Boy)	梅若の涙雨	うめわかのなみだ あめ	1784
Umō (Very Fine Rain with Tiny Droplets)	雨毛	うもう	1392
Umoku (Getting Wet from the Rain / Bathing Hair and Body in the Rainfall)	雨沐	うもく	1600
Umu (Fog-like Rain, Mist Rain)	雨霧	うむ	0948
Un'u (Clouds and Rain)	雲雨	うんう	1539
Un'u no Kokoro (Cloud, Rain, Heart, Soul, Spirit / A Graceful Heart That Calls for Clouds and Brings about the Rain)	雲雨の心	うんうのこころ	0970

Un'u no Majiwari (Fellowship of Cloud and Rain / Vows between a Man and Woman)	雲雨の交わり	うんうのまじわり	1738
Un'ufuzan (Cloud, Rain, Mountain—Parable about the Relationship between a Man and a Woman, the Goddess of the Mountain Said There Will Be a Cloud in the Morning and It Will Rain in the Evening)	雲雨巫山	うんうふざん	0933
Unkō Ushi (The Cloud Moves and It Rains in Various Places)	雲行雨施	うんこううし	1378
Unohana Kutashi (Rain That Rots the Rabbit Flowers / Long Rain before the Rainy Season)	卯の花腐し	うのはなくたし	1983
Unokoku Ame (Rabbit Time Rain / Rain in the Early Morning)	卯の刻雨	うのこくあめ	0135
Unokokuame (6 AM Rain)	卯刻雨	うのこくあめ	1996
Unotokiame (Rain That Starts at the Hour of the Rabbit, 5–7 AM)	卯時雨	うのときあめ	1633
Unpon Ufuku (If You Turn Your Palm Up, Clouds Will Form; If You Turn Your Palm Back, It Will Rain)	翻雲覆雨	うんぽんうふく	0219
Unraikuseiden Gōbakujudaiu Nenpikannonriki Ōjitokushōsan (There Are Days of Rain, Thunder, and Hail in This Life)	雲雷鼓掣電 降電 澍大雨 念彼観音 力 応時得消散	うんらいくせいでん ごうばくじゅだいう ねんぴかんのんき おうじとくしょうさん	1333
Uo (Getting Wet or Dirty from the Rain)	雨汙	うお	1305
Uraku (As the Rain Is Falling)	雨落	うらく	1304
Uranishi (Westerly or Northwesterly Wind with Rain That Blows from Late Autumn to Winter)	浦西	うらにし	1022

Ure Ame (Heavy Evening Rain—Western Shizuoka)		うれあめ	0934
Uri (Amid a Rainfall)	雨裏	うり	0957
Urin (Rain Forest)	雨林	うりん	1307
Urin Reikyoku (A Mourning Song Named after the Harmonious Sound of Rain and Bells on a Horse)	雨霖鈴曲	うりんれいきょく	1306
Urō (Rain and Dew)	雨露	うろ	0489
Urō (Rainwater or a Puddle Made by Rainfall)	雨潦	うろう	1310
Urō Sōsetsu (Rain, Dew, Frost, Snow / Changes in Various Weather or Various Sufferings and Worries)	雨露霜雪	うろそうせつ	0317
Uryō (The Amount of Rain That Falls on the Ground)	雨量	うりょう	1308
Uryōkei (Measurement of Rain)	雨量計	うりょうけい	1783
Uryoku Jurin (Rain Green Forest / A Forest That Develops during Monsoon)	雨緑樹林	うりょくじゅりん	0956
Uryū (Stand in the Rain)	雨立	うりゅう	1309
Uryū Ensa (Describes the Appearance of a Fisherman Working in the Rain)	雨笠煙蓑	うりゅうえんさ	0012
Usan (The Rain Vanishes)	雨散	うさん	1095
Usei (The Sound of Rain Falling)	雨声	うせい	1357
Usei (The Speed, Velocity, and Momentum of the Rain)	雨勢	うせい	1356
Usei (The Wind That Blows the Clouds after the Rain and Makes the Sky Clear)	雨晴	うせい	1729

Useki (Rain That Continues)	雨積	うせき	1098
Usen Fūma (Rain Washed and Wind Polished / The Mind and Spirit Are Cleansed, Strengthened, and Disciplined by Exposure to Rain and the Wind)	雨洗風磨	うせんふうま	1379
Usetsu (Rain and Snow)	雨雪	うせつ	0223
Ushi (God of Rain)	雨師	うし	0487
Ushi (Raining Far and Wide, Providing for All Living Things)	雨施	うし	1355
Ushi no Ame (Rain That Begins to Fall around the Day of the Ox or Hour of the Ox That Falls All Day Long)	丑の雨	うしのあめ	0027
Ushiame (Rain of the Hour of the Ox, between 1 and 3 AM, It Is Said That It Will Rain All Day Long)	丑雨	うしあめ	1353
Ushin Gesseki (Rain in the Morning and Evening When the Moon Rises)	雨晨月夕	うしんげっせき	1021
Ushiro Kara Zotto Suru Zo Yo Tsuyu-shigure ("From Behind Come Chilling Dewdrops like the Cold Autumn Rain" —Kobayashi Issa)	後から ぞっとす るぞよ 露時雨 (小林一茶)	うしろから ぞっとす るぞよ つゆしぐれ (こばやしいっさ)	0200
Ushitsu (To Be Wet and Clammy in the Rain)	雨湿	うしつ	1026
Ushō (Mountain Ranges with Pouring Rain)	雨嶂	うしょう	1096
Ushō (To Sing While Being Rained upon and Wet from the Rain)	雨嘯	うしょう	1354
Ushū (Collecting a Lot of Rain)	雨集	うしゅう	1782
Usō (By the Rainy Windowsill)	雨窓	うそう	1986

Usoku (Rain That Looks like White Threads Falling from the Sky)	雨足	うそく	1358
Usui (Pouring Rain)	雨水	うすい	1368
Usui (Rainwater, the Second Season in the Twenty-four Sekki Ancient Calendar/ The Season the Snow Changes into Rain and Ice Changes into Water)	雨水	うすい	0080
Utaku (Heavy and Intense Rain during the Season)	雨濯	うたく	1350
Utaku (Rain of Grace and Blessings from Heaven That Moistens All Things)	雨沢	うたく	1433
Uteki (Raindrop)	雨滴	うてき	0525
Utekisei (The Pattern of Raindrops Falling)	雨滴聲	うてきせい	1556
Utekisei (The Voice of a Raindrop)	雨滴声	うてきせい	0486
Utekisei ("The Voice of Rain-drops," from the Blue Cliff Record Compilation of Zen Buddhist Kōans published in 1125—Zengo)	雨滴聲（碧巌録）	うてきせい（へきがんろく）	0151
Uten (Raindrops)	雨点	うてん	1024
Uten (Rainy Weather)	雨天	うてん	0496
Uten Chūshi (Washed Out and Canceled by Rain)	雨天中止	うてんちゅうし	1769
Uten Enki (Postponed Because of the Rain)	雨天延期	うてんえんき	1351
Uten Junen (Rescheduled Due to the Rain)	雨天順延	うてんじゅんえん	1369
Uten Kekkō (To Continue Regardless of the Rain)	雨天決行	うてんけっこう	1770
Uten Tsuzuki (A Long and Continued Spell of Rainy Weather)	雨天続き	うてんつづき	0503

Utoūnabushinuami (Seasonal Rain—Hateruma Islands, Okinawa)		うとぅなぶしぬあみ	1023
Uun (Rain Cloud)	雨雲	ううん	0973
Uun no Kokoro (Heart like a Raincloud)	雨雲心	ううんのこころ	1799
Uun'noi (Rain Cloud Clothing / Getting Wet from the Rain)	雨雲衣	ううんのい	1676
Uya (Rainy Night)	雨夜	うや	0218
Uya no Tsuki (Rainy Night Moon)	雨夜の月	うやのつき	1534
Uyō (After the Rain)	雨余	うよ	0038
Uyō (The Rainy and the Sunny, the Highs and the Lows)	雨暘	うよう	1846
Uzen (Before the Rain)	雨前	うぜん	0707
Uzetsu (The Rain Ceases)	雨絶	うぜつ	1371
Uzui (When the Rain Follows Something as if It Were Trying to Say Something / Once the Rain Followed the Path of a Kind Governor during a Drought)	雨随	うずい	1346
Uzuki (Rain Month / The Lunar Month of May)	雨月	うづき	0195
Wafū Saiu (Calm and Kind Rain, Gentle Rain / Japanese Rain)	和風細雨	わふうさいう	0804
Waga Horishi Amewa Furikinu ("Long Wished-For Rain Has Fallen"—Man'yōshū [Collection of Ten Thousand Leaves], Volume 18, Song 4124)	我が欲りし雨は降り来ぬ（万葉集第十八巻 四千百二十四番歌）	わがほりしあめはふりきぬ（まんようしゅう だい18かん 4124ばんか）	1781
Wagamama Ame (Selfish Rain / Rain That Falls on Only One Part of an Area)	我儘雨	わがままあめ	0236

Waita (A Rainstorm That Blows Suddenly from the Northeast and Is Feared by Fishermen)		わいた	0705
Waiu (Long Rain / A Rain Long Enough to Affect Crop Harvesting)	淮雨	わいう	0484
Wakaba Ame (Rain Falling on New Leaves)	若葉雨	わかばあめ	0609
Watakushi Ame (Rain That Falls in an Isolated Area of the Mountain Even Though It Is Sunny Elsewhere)	私雨	わたくしあめ	0052
Wau (Japanese Rain / Good and Soft Rain for People and Crops)	和雨	わう	0105
Yabusame (Thicket Rain Bird)	藪雨	やぶさめ	1517
Yaeame (Rain Pouring Down in Layers and Layers)	八重雨	やえあめ	0706
Yakuu (November Medicine Rain That Hibernating Animals Drink in Preparation for Winter)	薬雨	やくう	1348
Yama No Ame (Mountain Rain)	山の雨	やまのあめ	0951
Yamaarashi (Mountain Storm)	山嵐	やまあらし	1621
Yamameguri (Rain in the Mountains)	山廻り	やまめぐり	0106
Yamashigure (Intermittent Rain in Winter Mountains)	山時雨	やましぐれ	0704
Yamashitazuyu (Dew Falling from the Branches and Leaves of Trees in the Mountain Forest)	山下露	やましたづゆ	1028
Yamaumushi (Rain in Early Spring—Shimane)	山蒸	やまうむし	1780
Yanagi no Ame (Willow Rain / Rain Falling on Willow Trees)	柳の雨	やなぎのあめ	0719

Yarai no Ame (Nonstop and Unrelenting Rain since the Night Before)	夜来の雨	やらいのあめ	0768
Yarai no Shigure (Sparse Night Rain / Intermittent Rain since Last Night)	夜来の時雨	やらいのしぐれ	0774
Yarazu no Ame (Rain That Stops Visitors from Returning Home)	遣らずの雨	やらずのあめ	0095
Yarehasu (Defeated Lotus— Signaling the Transition from Autumn to Winter, Even the Lotus That Withstood the Autumn Wind Has Been Torn by Cold Rain)	破れ蓮	やれはす	1844
Yasashii Ame (Calm Rain That Soothes One's Soul)	優しい雨	やさしいあめ	0805
Yashun (Evening Rain)	夜春	やしゅん	0806
Yau (Night Rain)	夜雨	やう	0189
Yau (Rain in the Plains)	野雨	やう	0739
Yaunissei (When It Rains at Night and Is Sunny during the Day / The Kind of Rain That the Farmers Like)	夜雨日晴	やうにっせい	0769
Yaunoki (Twilight Rain as Unexpectedly Appealing, Draws Attention to the Ephemerality of Existence)	夜雨の奇	やうのき	1349
Yautaishō (Night Rain and the Floor / A Good Sibling Relationship, Line Up the Beds on the Floor and Listen to the Rain at Night Together)	夜雨対牀	やうたいしょう	1365
Yaya tsuyoi Ame (Slightly Heavy Rain)	やや強い雨	ややつよいあめ	0102
Yayoshigure (A Drizzle That Falls Over and Over Again)	彌時雨	やよしぐれ	0414

Yo agari Tenki Ame Chikashi (It Will Rain If the Rain Starts in the Night)	夜上がり天気雨近し	よあがりてんきあめちかし	1722
Yo ni Furu Mo Sara Ni Sougi No Yadori Kana ("This World Is Just a Shelter from the Shower" —Matsuo Bashō)	世にふるも さらに宗祇の やどり哉（芭蕉）	よにふるも さらにそうぎの やどりかな（ばしょう）	1843
Yo no Mizore (Sleet Falling at Night)	夜の霙	よのみぞれ	0803
Yo no Shigure (A Quickly Passing Shower in the Night)	夜の時雨	よのしぐれ	1364
Yoagari (It Rains at Night and Then Stops at Night)	夜上	よあがり	0771
Yoarashi (Night Storm)	夜嵐	よあらし	0018
Yobe no Ame (Last Night's Rain / When One Wakes and Sees Leaves Carrying the Rain from the Night Before)	昨夜の雨	よべのあめ	0104
Yobidoi (Rainspout)	呼びどい	よびどい	0833
Yodachi (To Depart at Night)	夜立ち	よだち	0770
Yoi no Ame (Sunset Rain)	宵の雨	よいのあめ	1497
Yōjyō Mushū no Ame ("Melancholy of Rain Season Droplets on Leaves"—Matsuo Bashō)	葉上無愁雨（芭蕉）	ようじょう むしゅうのあめ（ばしょう）	1367
Yōka no Ame (Rain Falling on the Cherry Blossoms High atop the Cold Mountains That Do Not Bloom until Early Summer)	余花の雨	よかのあめ	0264
Yōka no Ame (Rain That Nourishes Flowers)	養花の雨	ようかのあめ	1498
Yōka-u (Rain That Encourages Flowers to Bloom / Spring Rain)	養花雨	ようかう	0501
Yokoame (Rain That Falls Sideways, Being Blown by the Wind)	横雨	よこあめ	0969

Yokoburi (It Rains from the Side Due to Strong Winds / Sideways Rain)	横降り	よこぶり	0133
Yokoita ni Amadare (Raindrop on a Board / Halting Speech)	横板に雨垂れ	よこいたにあまだれ	1842
Yokonaguri (Rain Striking from the Side)	横なぐり	よこなぐり	1372
Yokonaguri no Ame (Driving, Slanted Rain)	横なぐりの雨	よこなぐりのあめ	0169
Yokoshigure (Rain That Falls Sideways in Winter)	横時雨	よこしぐれ	0127
Yokozamaame (Sideways Rain and Wind)	横方雨	よこざまあめ	0974
Yokurin (Rain That Fertilizes Farmland)	沃霖	よくりん	0939
Yokuu (Rain with a Lot of Nutrients)	沃雨	よくう	0940
Yomake (Rain That Stops at Nightfall—Chita Region, Aichi)		よまけ	1366
Yomenakase Biyori (Rain That Falls at Night and Turns to Sunshine during the Day)	嫁泣かせ日和	よめなかせびより	0936
Yoreki (Lingering Raindrops)	余瀝	よれき	0061
Yoru No Ame (Rain That Falls at Night)	夜の雨	よるのあめ	0766
Yosame (Rain That Falls at Night)	夜雨	よさめ	0758
Yōseibaiu (Whimsical Rain That Is Changeable and Could Be Raining Buckets One Moment and Sunny the Next)	陽性梅雨	ようせいばいう	1841
Yoteki (Remaining Drops of Rain)	余滴	よてき	1007

Youu Ame o Kīte Kankō Tsuki Mon o Hirakeba Rakuyō Ōshi (A Cold Night Passes and Becomes Dawn While Listening to the Rain; When You Open the Gate, Leaves on the Ground, It was Not Rain but the Sound of Falling Leaves on the Roof)	葉雨聴雨寒更尽開門落葉多	ようう あめをきいてかんこうつき もんをひらけば らくようおおし	0178
Yowa no Ame (Twilight Rain That Falls Deep into the Night)	夜半の雨	よわのあめ	0757
Yowai Ame (Weak Rain)	弱い雨	よわいあめ	1612
Yowai Yuki (Soft Snow or Rain)	弱い雪	よわいゆき	0103
Yōzu (Early Spring Rain That Melts the Winter Snow)		よーず	0968
Yōzu (Hot and Humid Rain Season)		よーず	1775
Yūdachi (Sudden Summer Rain with Thunder in the Evening)	夕立	ゆうだち	0807
Yūdachi wa Mikka (When There Is a Sudden Evening Summer Rain Shower, It Will Rain for Three Days)	夕立は三日	ゆうだちはみっか	0435
Yūdachi ya Ta o Mimeguri no Kami Naraba ("If You Are Truly a Deity Watching over Rice Fields, Let It Rain by Evening"—Successful Rain Prayer Poem during Drought by Takai Kikaku, Mimeguri Shrine)	夕立や 田を見めぐりの 神ならば（宝井其角）	ゆうだちや たをみめぐりの かみならば（たからいきかく）	0039
Yūdachiame (A Sudden Summer Rain)	夕立雨	ゆうだちあめ	1845
Yūdachibare (Sun Will Appear after the Afternoon Rain / Summer)	夕立晴	ゆうだちばれ	0813
Yūdachigumo (Rain Clouds in Summer)	夕立雲	ゆうだちぐも	0810

Yūdatsu (When It Showers on a Spring Afternoon)	夕立つ	ゆうだつ	0938
Yudun (Rainy Season— Ishigakijima Region, Okinawa)		ゆどぅん	0010
Yueiu (Spring Rain)	愉英雨	ゆえいう	0937
Yufudachi (Raining in the Evening)	夕立	ゆふだち	0809
Yufudatsu (Occurring in the Evening / Rain in the Evening)	夕立つ	ゆふだつ	0808
Yūfūshiu (The Cloud Befriends the Wind and Bears Rain as Their Child)	友風子雨	ゆうふうしう	1736
Yūhi ga Mabushikunai no wa, Ame ga Furu Shirushi (If the Sunset Is Not Bright and Brilliant, Rain)	夕日がまぶしくないのは、雨が降るしるし	ゆうひがまぶしくないのは、あめがふるしるし	1121
Yukiame (Snow Rain / Rain Mixed with Snow)	雪雨	ゆきあめ	1006
Yukigeame (Rain in Early Spring That Melts the Snow)	雪解雨	ゆきげあめ	0504
Yukigeame (Spring Rain That Melts the Snow)	雪解雨	ゆきげあめ	0340
Yukikeshi no Ame (Rain That Erases Snow)	雪消しの雨	ゆきけしのあめ	0255
Yukimajiri (Rain and Wind Mixed with Snow)	雪交じり	ゆきまじり	1005
Yukimajiri (Snow Mixed with Rain and Wind)	雪雑り	ゆきまじり	0720
Yukioroshi (Cold Rain with Strong Wind and Lightning, Signaling the Arrival of Winter)	雪嵐	ゆきおろし	0815
Yukishigure (Rain Mixed with Sleet and Snow)	雪時雨	ゆきしぐれ	0502
Yukyou no Ame (Rain Falling on Elm Flower Petals)	楡莢の雨	ゆきょうのあめ	1455

Yūmizore (Evening Sleet)	夕霙	ゆうみぞれ	1119
Yūniji (Evening Rainbow)	夕虹	ゆうにじ	1454
Yūsame (Evening Rain)	夕雨	ゆうさめ	1457
Yūsetsu (The Snow Melts from Heavy Rain Falling)	融雪	ゆうせつ	0338
Yūshigure (Winter Drizzle in the Evening)	夕時雨	ゆうしぐれ	0488
Yuta (Rain-Strike Roof in Buddhist Temple Architecture)	雨打	ゆた	0500
Zabun (Sound of Rain)		ざぶん	0302
Zaburi (Heavy Rain That Falls Repeatedly in the Same Day)		ざぶり	1838
Zabuzabu (Sound of Rain)		ざぶざぶ	0411
Zan'nu (A Faint Rain That Remains and Lingers after a Stronger Rain)	残雨	ざんう	1120
Zanteki (Remaining Raindrops after Rain)	残滴	ざんてき	1893
Zanzaburi (Very Heavy Downpour of Rain)		ざんざぶり	0812
Zanzan (Sound of Rain Falling in Big Drops / Sloshing Rain)		ざんざん	0329
Zāzā (The Sound of Heavy Rain)		ざあざあ	0305
Zāzāfuri (A Drenching Downpour of a Very Heavy Rain / The Sound That Heavy Rain Makes)	ザーザー降り	ざーざーふり	0464
Zen'nyo (Dragon God Who Lives in a Pond and Is Prayed to for Rain)	善如	ぜんにょ	1604
Zen'nyo Ryūō (Female Rain Dragon Deity Whom Kūkai Famously Made Appear during a Rainmaking Contest in 824 at the Kyoto Imperial Palace)	善女龍王	ぜんにょりゅうおう	0649

Zensensei Kōu (Frontal Rain)	前線性降雨	ぜんせんせいこうう	1839
Zentatsu (Rain Deity Dragon Who Lived on Murōyama Mountain, Nara Prefecture)	善達	ぜんたつ	1334
Zubunure (Soaked by Rain All the Way through One's Clothing)	ずぶ濡れ	ずぶぬれ	1840
Zui-u (A Welcome Rain That Encourages Grain to Grow / A Rain of Grace)	瑞雨	ずいう	0319

ACKNOWLEDGMENTS

I wish to express my deep gratitude to the many people who have supported this endeavor. Thank you first and foremost to the MIT Press and to my editor Gabriela Bueno Gibbs for trusting in this project. To my kind husband, Shelter Serra, who patiently listened to me describe various types of rain for three years, and my parents, Michiko and Richard Miller, for their unwavering faith in all of my pursuits. Lorraine Young for her impeccable editing and friendship. Liam van Loenen, Patricia Moritz, and Daniel Kelm for their belief in the project from the outset. Jess Spector, my right-hand man, whose equanimity kept my studio running harmoniously. Susan McCaffrey, whose hard work allowed me the time to write instead of paint. Hanae Kawai, Kumi Kishida, Xin Lian, Pavlo Terekhov, and Tiffany Chung, without whose assistance this compendium would not have been possible. Serena Caffrey for the thoughtful words, which always gave me solace. Thank you to the contributors for their profoundly beautiful insights and contextualization of this work. Finally, this work was created in the memory of my grandparents, Buddhist Priest Ando Gakujyō Shonin and Ando Miyako, without whom I would never have noticed all of the poetry and subtleties in the natural world. ありがとう ございます.

NOTES

FOREWORD

1. Rein Raud, "The Existential Moment: Re-reading Dōgen's Theory of Time," *Philosophy East and West* 62, no. 2 (April 2012), accessed May 27, 2024, https://en.wikipedia.org/wiki/D%C5%8Dgen#cite_ref-52.

2. Described as "the facticity of all things" by Yuriko Saito, "Historical Overview of Japanese Aesthetics," in *Introduction to New Essays in Japanese Aesthetics*, ed. A. Minh Nguyen (Lanham, MD: Lexington Books, 2017), digital book, unpaginated.

3. Kamo no Chomei, "An Account of My Hut," in "Medieval—Kamo no Chomei (1153–1216)," *Five Japanese Authors*, chapter 3, accessed June 4, 2024, https://www.washburn.edu/reference/bridge24/Hojoki.html. Dwelling places were of particular importance to Chōmei as so many were destroyed in Kyoto at the time, and he did not find a long-term place of residence until he built his portable ten-square-foot hut.

4. All quotations from Miya Ando are drawn from a conversation with the author, May 16, 2024.

5. Sonja Arntzen, *Kagerō Diary: A Woman's Autobiographical Text from Tenth-Century Japan* (Ann Arbor: University of Michigan Center for Japanese Studies, 1998), https://www.fulcrum.org/epubs/ww72bd40v?locale=en#/6/22[nav_10]!/4/150/2/2/6/2/1:0, location 112; https://doi.org/10.3998/mpub.18535. Womer poets of the Heian period, such as Michitsuna no Haha, were the first to add native sensibility to this art form after centuries of Chinese influence.

6. Described as "the facticity of all things" by Saito, "Historical Overview of Japanese Aesthetics," n.p.

7. Steve Odin, "Artistic Detachment in Japanese Aesthetics," in *Artistic Detachment in Japan and the West: Psychic Distance in Comparative Aesthetics* (Honolulu: University of Hawai'i Press, 2001), 99.

8. Saito, "Historical Overview of Japanese Aesthetics," n.p.

9. Dennis Hirota, *Wind in the Pines: Classic Writings of the Way of Tea as a Buddhist Path* (Kyoto: Ryokoku University and Dennis Hirota, 1995), 43–44.

10. *Senzaishū* anthology, vol. 6, no. 414 (translated and posted November 14, 2015), WakaPoetry.net, accessed June 2, 2024, https://www.wakapoetry.net/szs-vi-414/.

11. Donald Keene, *Essays in Idleness: The Tsurezuregusa of Kenkō* (Tokyo: Tuttle Publishers, 1989), 115.

12. A continuum in Japanese painting aesthetics that Ando holds to in the *Water of the Sky* series is that of going beyond lifelikeness, not precisely copying the superficial form of her subject, but evoking the qualities of its natural energies. This aesthetic mandate was put into writing by Tosa Mitsuoki (Japan, 1617–1691) in his *Authoritative Summary of the Rules of Japanese Painting* (*Honchō gahō taiden*), in which he terms proper lifelikeness as following the "laws of life," and superior artworks that go beyond the lifelike as following the "laws of painting." See Makoto Ueda, *Literary and Art Theories in Japan* (Cleveland: Press of Case Western Reserve University, 1967), 130.

PREFACE

1. This pre-Gregorian early modern Japanese calendar, known as the Pentads calendar, describes seventy-two seasons defined by local phenological events. The seventy-two-seasons, or *Kō*, Pentads calendar 七十二候 is an adaptation of an older, twenty-four-part Chinese Solar Terms (*Sekki*) 二十四節気 calendar, which originated in the Chinese Yellow River basin circa 2,500 years ago. The Solar Terms calendar is based upon the sun's ecliptic orbit; every fifteen degrees (which, in turn, correspond to fifteen days) signifies a new season in this system. When this calendar was imported to Japan several hundred years ago, it was

further divided into seventy-two pentads, three pentads per solar term at approximately five days per season, with the names of each season also being customized to the Japanese climate and terrain.

2. Adapted from Liam van Loenen, "Ama No Mizu (Water of the Sky / Water of the Heavens)," unpublished essay, April 2022.

3. The full set of 2,000 drawings is archived on the Rain Dictionary website, www.raindictionary.com.

4. The system of Hepburn romanization or Hebon-Shiki (ヘボン式) has been employed in this project.

5. 天 is a term that can be read as *ama*, *ame*, or *ten*; it refers to "the above" or the heavenly, spiritual, celestial, and metaphysical realms. Archaic and poetic terms for rain include *ama no mizu*, *ame no mizu*, and *ten no mizu* (天の水).

6. *Man'yōshū* or *Collection of Ten Thousand Leaves* is an anthology of ancient Japanese poems compiled circa 759 CE during the Nara Period but including many earlier works.

7. In the Rain Dictionary website, see drawing numbers 0131 (Sowofuru [Elegant and Quiet Rain]), 0145 (Ame Sōburu [Rain Falls Slowly—Ise Monogatari, Second Stage]), 0176 (Sobofuru [Steady Rain]), 0273 (Sohoburu [Gentle Rain]), 0287 (Sōburu [Silent and Gentle Rain]), 0343 (Sohoburu [Rain That Falls Quietly and Gently]), 0388 (Sowoburu [Elegant Rain]), 0434 (Sohotzu [To Become Wet with Rain or Tears]), 0794 (Sobotsu [Rain That Falls Elegantly and Steadily]), 0836 (Sohofuru ["It Rains Gracefully"—Song from the Country of Koshinomichi No Naka, *Man'yōshū* [*Collection of Ten Thousand Leaves*], Volume 16, Song 3883]), 1154 (Sobofuruame [Smooth and Calm Rain]), 1187 (Sohoburu [Graceful and Slow Soft Rain—Ise Monogatari]), 1257 (Sohotsu [Rain or Tears That Fall Gracefully]), 1325 (Amesobofuru [Tasteful Rain That Falls Steadily, Quietly, Gracefully]), 1541 (Sohofuru [It Rains Slowly]), 1607 (Sohoburu [Soft Slow Rain]), and 1972 (Sōtsu [Raining Gracefully]).

8. Rain, too, is a keeper of time, as described in drawing numbers 0172 (Amaochi Byōshi [Rhythm That Mimics the Raindrops Falling from the Roof, Used When Learning Shamisen]) and 0557 (Amadare Chōshi [Raindrops' Rhythm / In Music an Inexperienced Person Playing Falteringly]), which both reference rain's metronome-like quality.

9. See drawing numbers 0114 Kitsune no Yometori Ame (Fox Taking a Wife Rain / Light Rain in Sunshine), 0126 Kitsune no Yomeiri (Rain That Falls Even Though the Sun Is Shining / The Day That Foxes Have Their Wedding Ceremony), 0148 Kitsune no Ame (Fox Rain / Rain That Seemingly Comes from Nowhere as if a Trick by a Fox Spirit), 0303 Tenkiame (It's Sunny But Raining / Fox Rain), 0307 Madowasareru Ame (Mysterious and Deceiving Rain, Possibly the Doing of a Ghosting Fox), 0962 Hideriame (To Have Rain While the Sun Is Shining / Wedding Day Celebration of Two Foxes Getting Married), 0975 Sobae (To Rain While the Sun Is Shining / The Marriage of a Fox / Summer), 1386 Terifuri (Fox Bride Rain / Raining While Sunny), 1513 Kitsunebiyori (Fox Weather / Raining and Sunny), 1531 Kitsuneame (Fox Rain / Light Rain on a Sunny Day), 1852 Kitsune no Goshūgi (Fox Celebration / Raining While Sunny, the Weather in Which the Foxes Have Wedding Ceremonies), 1979 Tirabui (Fox Bride Rain / Rain in the Sunshine).

10. Ame o Kiku (drawing 0816).

11. Ame No Oto Wo Kokoro Ni Shimiiru Yōni Kiku (drawing 2048 in an in-progress addendum to the Rain Dictionary).

PERCEIVING RAIN

1. Thomas Merton, "Rain and the Rhinoceros," in *Raids on the Unspeakable* (1960; repr., New York: New Directions, 1964), 10.

2. Dōgen, "Mountains and Waters Sutra," translated by Kazuaki Tanahashi, https://www.upaya.org/uploads/pdfs/MountainsRiversSutra.pdf.